Plant Growth and Development

TEACHER'S GUIDE

SCIENCE AND TECHNOLOGY FOR CHILDREN

NATIONAL SCIENCE RESOURCES CENTER
Smithsonian Institution–National Academy of Sciences
Arts and Industries Building, Room 1201
Washington, DC 20560

NSRC

The National Science Resources Center is operated by the National Academy of Sciences and the Smithsonian Institution to improve the teaching of science in the nation's schools. The NSRC collects and disseminates information about exemplary teaching resources, develops and disseminates curriculum materials, and sponsors outreach activities, specifically in the areas of leadership development and technical assistance, to help school districts develop and sustain hands-on science programs. The NSRC is located in the Arts and Industries Building of the Smithsonian Institution in Washington, D.C.

STC Project Supporters

National Science Foundation
Smithsonian Institution
U.S. Department of Defense
U.S. Department of Education
John D. and Catherine T. MacArthur Foundation
The Dow Chemical Company Foundation
E. I. du Pont de Nemours & Company
Amoco Foundation, Inc.
Hewlett-Packard Company
Smithsonian Institution Educational Outreach Fund

This project was supported, in part,
by the
National Science Foundation
Opinions expressed are those of the authors
and not necessarily those of the Foundation

ISBN 0-89278-633-7

Published by Carolina Biological Supply Company, 2700 York Road, Burlington, NC 27215.
Call toll free 800-334-5551.

This material is based upon work supported by the National Science Foundation under Grant No. ESI-9252947. Any opinions, findings, and conclusions or recommendations expressed in this material are those of the author(s) and do not necessarily reflect the views of the National Science Foundation.

CB787049710

♻ Printed on recycled paper.

Foreword

Since 1988, the National Science Resources Center (NSRC) has been developing Science and Technology for Children (STC), an innovative hands-on science program for children in grades one through six. The 24 units of the STC program, four for each grade level, are designed to provide all students with stimulating experiences in the life, earth, and physical sciences and technology while simultaneously developing their critical thinking and problem-solving skills.

Sequence of STC Units

Grade	Life, Earth, and Physical Sciences and Technology			
1	Organisms	Weather	Solids and Liquids	Comparing and Measuring
2	The Life Cycle of Butterflies	Soils	Changes	Balancing and Weighing
3	Plant Growth and Development	Rocks and Minerals	Chemical Tests	Sound
4	Animal Studies	Land and Water	Electric Circuits	Motion and Design
5	Microworlds	Ecosystems	Food Chemistry	Floating and Sinking
6	Experiments with Plants	Measuring Time	Magnets and Motors	The Technology of Paper

The STC units provide children with the opportunity to learn age-appropriate concepts and skills and to acquire scientific attitudes and habits of mind. In the primary grades, children begin their study of science by observing, measuring, and identifying properties. Then they move on through a progression of experiences that culminate in grade six with the design of controlled experiments.

Sequence of Development of Scientific Reasoning Skills

Scientific Reasoning Skills	Grades					
	1	2	3	4	5	6
Observing, Measuring, and Identifying Properties	♦	♦	♦	♦	♦	♦
Seeking Evidence Recognizing Patterns and Cycles		♦	♦	♦	♦	♦
Identifying Cause and Effect Extending the Senses				♦	♦	♦
Designing and Conducting Controlled Experiments						♦

The "Focus–Explore–Reflect–Apply" learning cycle incorporated into the STC units is based on research findings about children's learning. These findings indicate that knowledge is actively constructed by each learner and that children learn science best in a hands-on experimental environment where they can make their own discoveries. The steps of the learning cycle are as follows:

- Focus: Explore and clarify the ideas that children already have about the topic.

- Explore: Enable children to engage in hands-on explorations of the objects, organisms, and science phenomena to be investigated.

- Reflect: Encourage children to discuss their observations and to reconcile their ideas.

- Apply: Help children discuss and apply their new ideas in new situations.

The learning cycle in STC units gives students opportunities to develop increased understanding of important scientific concepts and to develop better attitudes toward science.

The STC units provide teachers with a variety of strategies with which to assess student learning. The STC units also offer teachers opportunities to link the teaching of science with the development of skills in mathematics, language arts, and social studies. In addition, the STC units encourage the use of cooperative learning to help students develop the valuable skill of working together.

In the extensive research and development process used with all STC units, scientists and educators, including experienced elementary school teachers, act as consultants to teacher-developers, who research, trial teach, and write the units. The process begins with the developer researching the unit's content and pedagogy. Then, before writing the unit, the developer trial teaches lessons in public school classrooms in the metropolitan Washington, D.C., area. Once a unit is written, the NSRC evaluates its effectiveness with children by nationally field-testing it in ethnically diverse urban, rural, and suburban public schools. At the field-testing stage, the assessment sections in each unit are also evaluated by the Program Evaluation and Research Group of Lesley College, located in Cambridge, Mass. The final editions of the units reflect the incorporation of teacher and student field-test feedback and of comments on accuracy and soundness from the leading scientists and science educators who serve on the STC Advisory Panel.

The STC project would not have been possible without the generous support of numerous federal agencies, private foundations, and corporations. Supporters include the National Science Foundation, the Smithsonian Institution, the U.S. Department of Defense, the U.S. Department of Education, the John D. and Catherine T. MacArthur Foundation, the Dow Chemical Company Foundation, the Amoco Foundation, Inc., E. I. du Pont de Nemours & Company, the Hewlett-Packard Company, and the Smithsonian Institution Educational Outreach Fund.

Acknowledgments

Plant Growth and Development was researched and developed by Patricia McGlashan. She worked closely with Paul Williams, professor of plant pathology at the University of Wisconsin, who developed the Wisconsin Fast Plant™, along with Coe Williams and Jane Sharer, coordinators of the Wisconsin Fast Plants Project. The unit was edited by Marilyn Fenichel and illustrated by Max-Karl Winkler. Other NSRC staff who contributed to the development and production of this unit include Sally Goetz Shuler, deputy director; Joe Griffith, STC project director; Kathleen Johnston, publications director; and Timothy Falb, publications technology specialist. The unit was evaluated by George Hein and Sabra Price, Program Evaluation and Research Group, Lesley College. *Plant Growth and Development* was trial taught at Watkins School of the Capitol Hill Cluster Schools in Washington, D.C.

The NSRC would like to thank the following individuals for their contributions to the unit:

Deborah Bevan, Teacher, Glendale Elementary School, Madison, WI

Anne Gallagher, Teacher, Carole Highlands School, Takoma Park, MD

Ann Gay, Principal, Capitol Hill Cluster Schools, Washington, DC

Barbara Giertz, Teacher, Carole Highlands School, Takoma Park, MD

Jean Gregor, Teacher, Carole Highlands School, Takoma Park, MD

Gene Hickey, Elementary Science Resource Teacher, Garfield Math/Science Elementary School, Milwaukee, WI

Audrey Humphries, Teacher, Watkins School, Capitol Hill Cluster Schools, Washington, DC

Veola Jackson, Principal, Watkins School, Capitol Hill Cluster Schools. Washington, DC

Debbie Johnson, Teacher, Carole Highlands School, Takoma Park, MD

Ulysses Johnson, Vice Principal, Watkins School, Capitol Hill Cluster Schools, Washington, DC

Wesley Licht, Teacher, Glendale Elementary School, Madison, WI

Dane Penland, Staff Photographer, Smithsonian Institution, Washington, DC

Barbara Scott, Principal, Carole Highlands School, Takoma Park, MD

Becky Smith, Curriculum Editor, Mesa Public Schools, Mesa, AZ

Catherine Taylor, Elementary Science Resource Teacher, Watkins School, Capitol Hill Cluster Schools, Washington, DC

Emma Walton, Science Program Coordinator, Anchorage School District, Anchorage, AK

Robert Zeman, Inservice Specialist/Science Kit Program, The Kluge Science Center, Milwaukee, WI

The NSRC would also like to thank the following individuals and school systems for their assistance with the national field-testing of the unit:

Gerry Consuegra, Elementary Science Coordinator, Montgomery County Public Schools, Rockville, MD

Geraldine Cutler, Teacher, Bunker Hill School, Washington, DC

Diane Hancock, Teacher, Pleasant Valley School, Portland, OR

Nancy Jones, Teacher, Garfield Elementary School, Spokane, WA

Dorothy Landis, Teacher, Somerset School, Chevy Chase, MD

Richard McQueen, Specialist, Science Education, Multnomah Education Service District, Portland, OR

Kathy Miller, Teacher, Linwood Elementary School, Spokane, WA

Carolyn Preston, Bunker Hill School, Washington, DC

Brenda Puglevand, Teacher, Bemiss Elementary School, Spokane, WA

Sue Rasmussen, Teacher, Menlo Park School, Portland, OR

Maryagnes Sisti, Teacher, Somerset School, Chevy Chase, MD

Scott Stowell, Curriculum Coordinator, Spokane Public School District 81, Spokane, WA

Sandy Wilber, Teacher, Menlo Park School, Portland, OR

Douglas Lapp
Executive Director
National Science Resources Center

STC Advisory Panel

Peter P. Afflerbach, Professor, National Reading Research Center, University of Maryland, College Park, MD

David Babcock, Director, Board of Cooperative Educational Services, Second Supervisory District, Monroe-Orleans Counties, Spencerport, NY

Judi Backman, Math/Science Coordinator, Highline Public Schools, Seattle, WA

Albert V. Baez, President, Vivamos Mejor/USA, Greenbrae, CA

Andrew R. Barron, Professor of Chemistry and Material Science, Department of Chemistry, Rice University, Houston, TX

DeAnna Banks Beane, Project Director, YouthALIVE, Association of Science-Technology Centers, Washington, DC

Audrey Champagne, Professor of Chemistry and Education, and Chair, Educational Theory and Practice, School of Education, State University of New York at Albany, Albany, NY

Sally Crissman, Faculty Member, Lower School, Shady Hill School, Cambridge, MA

Gregory Crosby, National Program Leader, U.S. Department of Agriculture Extension Service/4-H, Washington, DC

JoAnn E. DeMaria, Teacher, Hutchison Elementary School, Herndon, VA

Hubert M. Dyasi, Director, The Workshop Center, City College School of Education (The City University of New York), New York, NY

Timothy H. Goldsmith, Professor of Biology, Yale University, New Haven, CT

Patricia Jacobberger Jellison, Geologist, National Air and Space Museum, Smithsonian Institution, Washington, DC

Patricia Lauber, Author, Weston, CT

John Layman, Director, Science Teaching Center, and Professor, Departments of Education and Physics, University of Maryland, College Park, MD

Sally Love, Museum Specialist, National Museum of Natural History, Smithsonian Institution, Washington, DC

Phyllis R. Marcuccio, Associate Executive Director for Publications, National Science Teachers Association, Arlington, VA

Lynn Margulis, Distinguished University Professor, Department of Botany, University of Massachusetts, Amherst, MA

Margo A. Mastropieri, Co-Director, Mainstreaming Handicapped Students in Science Project, Purdue University, West Lafayette, IN

Richard McQueen, Teacher/Learning Manager, Alpha High School, Gresham, OR

Alan Mehler, Professor, Department of Biochemistry and Molecular Science, College of Medicine, Howard University, Washington, DC

Philip Morrison, Professor of Physics Emeritus, Massachusetts Institute of Technology, Cambridge, MA

Phylis Morrison, Educational Consultant, Cambridge, MA

Fran Nankin, Editor, *SuperScience Red*, Scholastic, New York, NY

Harold Pratt, Senior Program Officer, Development of National Science Education Standards Project, National Academy of Sciences, Washington, DC

Wayne E. Ransom, Program Director, Informal Science Education Program, National Science Foundation, Washington, DC

David Reuther, Editor-in-Chief and Senior Vice President, William Morrow Books, New York, NY

Robert Ridky, Professor, Department of Geology, University of Maryland, College Park, MD

F. James Rutherford, Chief Education Officer and Director, Project 2061, American Association for the Advancement of Science, Washington, DC

David Savage, Assistant Principal, Rolling Terrace Elementary School, Montgomery County Public Schools, Rockville, MD

Thomas E. Scruggs, Co-Director, Mainstreaming Handicapped Students in Science Project, Purdue University, West Lafayette, IN

Larry Small, Science/Health Coordinator, Schaumburg School District 54, Schaumburg, IL

Michelle Smith, Publications Director, Office of Elementary and Secondary Education, Smithsonian Institution, Washington, DC

Susan Sprague, Director of Science and Social Studies, Mesa Public Schools, Mesa, AZ

Arthur Sussman, Director, Far West Regional Consortium for Science and Mathematics, Far West Laboratory, San Francisco, CA

Emma Walton, Program Director, Presidential Awards, National Science Foundation, Washington, DC, and Past President, National Science Supervisors Association

Paul H. Williams, Director, Center for Biology Education, and Professor, Department of Plant Pathology, University of Wisconsin, Madison, WI

Contents

Goals for *Plant Growth and Development*

In this unit, students observe the life cycle of the *Brassica rapa* (Wisconsin Fast Plants™). Their experiences introduce them to the following concepts, skills and attitudes.

Concepts

- Many plants follow a life cycle that begins with growth from a seed and proceeds through the production of seeds.

- Plants have distinct stages in their life cycle.

- To live and grow, plants need light, water and nutrients from the soil.

- Flowering plants must be pollinated in order to produce seeds.

- Many plants are pollinated by bees.

- A flower's pollen sticks to a bee, but some rubs off when the bee feeds at other flowers.

- One seed produces one plant; one plant can produce many seeds.

Skills

- Planting and caring for *Brassica rapa*.

- Observing, describing, and recording changes in plants.

- Comparing and discussing changes occurring in plants over time.

- Measuring and recording the growth of plants.

- Using graphs to display and compare growth patterns.

- Predicting future growth from observations and measurements.

- Reading to learn more about plants.

- Communicating results and reflecting on experiences through writing, drawing and discussion.

Attitudes

- Developing interest in studying the life cycle of plants.

- Developing sensitivity to the needs of plants.

- Developing an awareness of the interaction between plants and animals.

Unit Overview and Materials List

Plant Growth and Development is an eight-week unit during which students experience the complete life cycle of a plant in a very short time and learn that the cycle includes germination, growth, development of specialized parts, and even death, with the promise of new life in the seed. The unit was designed for grade 3 but also could be taught at grade 4. Extensive classroom testing of the unit showed that third graders were highly interested in the unit, displayed the necessary manual dexterity, and could comprehend the major concepts introduced.

The unit features rapid-cycling Wisconsin Fast Plants™, which go from seed to seed in 40 days. Wisconsin Fast Plants are *Brassicas* (the mustard and cabbage family), and were developed over a period of 15 years by Dr. Paul Williams of the University of Wisconsin.

The unit opens with lessons on observing seeds and brainstorming about what the students already know about plants. These lessons are followed by a planting activity that stresses following directions and working independently. In about 24 hours, students can observe the seedlings emerge and begin to record their observations both in writing and by making scientific drawings. Several days later, the students will gain experience with two practical gardening techniques, thinning and transplanting, and learn when they should be used.

The unit emphasis then shifts to the theme of interdependence and explores the reasons why the bee and the flower need each other. Since interdependence is such an important concept, students work with it in several different ways.

They cross-pollinate their own plants using real bees on toothpicks. Based on their observations, they construct models of bees and blossoms, and act out pollination using their models.

Throughout the unit, students are encouraged to make frequent observations of their plants using as much sensory information as possible. Students continue to record these observations in writing and drawing. They also quantify their observations by taking frequent measurements and recording these on growth graphs.

Finally, the students harvest and thresh the "crop" and determine their yield. There are suggestions of what to do with the seed to extend the unit into more individualized experiments.

To help make management of the unit easier, special instructions about setting up the equipment and other issues have been flagged with icons. Watch for the icon—a finger tied with a bow—for important information.

The **Appendices** contain a number of useful items. The first section, **Post-Unit Assessments**, provides tools to help you evaluate student progress in understanding the plant's life cycle and learning bee anatomy. The second section contains a complete lesson on graphing, in case your class has not yet studied graphing or could use a review before starting the unit. The unit assumes that students can graph independently, so consider teaching this lesson some time before planting day. There also are black-line masters for reproducing graph paper and observation sheets, life cycle cards, and an annotated bibliography of books for both students and teachers.

You do not have to be an expert in botany to teach this unit. The background sections of the Teacher's Guide will provide you with most of the information you need. But don't be surprised if you find yourself learning along with the students, and if you and your students find yourselves faced with puzzling questions. Use this situation to model the way scientists learn: define the question, then ask, "How can we find out?" This will encourage your students to find their own answers by experimenting and consulting resource materials.

Materials List

Below is a list of materials needed for the *Plant Growth and Development* unit.

1		Teacher's Guide
15		Student Activity Books
30		trays
1	pkg.	bean seeds (lima beans work well)
1	pkg.	toothpicks
30		spoons
15		droppers
30		paper cups
60		plastic cups
1	pkg.	dried honeybees
6		large paper cups
30		pairs of forceps
30		hand lenses
2	pkgs.	Wisconsin Fast Plants™ seeds
2	pkgs.	fertilizer
2	pkgs.	potting mix
2	pkgs.	wooden stakes
2	pkgs.	plastic rings
30		planter labels
30		planter quads
3		water tanks
3		water mats
120		wicks
3		felt squares containing copper sulfate (see pgs. 5 and 6)
1		lighting system
1		bag of 500 snap-together centimeter cubes

*		Student notebooks
*		Drawing paper
*	1	large jar
*		24" X 36" newsprint pad and markers
		OR
*		Overhead transparencies and markers, with projector and screen
*		Paper towels
*		Sponges
*	2	dish pans
*		Whisk broom and dustpan
*30		pairs of scissors
*		Glue and small glue cups
*		Crayons
*		Supplies for making flower models (see Lesson 13)
*		Supplies for making bee models (see Lesson 14)

*Note: These items are not included in the kit. Including them would increase material and shipping costs, and they are commonly available in most schools or can be brought from home.

Timetable for Wisconsin Fast Plants™

Lesson No.	Activity	Date (Day numbers)
	Preparation	
	Assemble lights	
	Set up watering system	
1, 2	Introductory activities	
3	Plant the seed (on a Monday or Tuesday)	1
	Water from top	1 2 3
4	Thin and transplant	4 5
	Check water level	5
5	Begin Plant Growth Graph	5 6 7 8
6	Observe true leaves and flower buds develop	7 8 9
7	Observe growth spurt	8 9 10 11 12 13 14 15
8	Draw bees	8 9
9	Make bee stick	10 11
	Check water level	12
10	Observe flowers open	12 13 14 15 16 17 18
11	Pollinate	12 13 14 15 16 17 18
	Pinch off unopened buds	18
	Check water level	19
12	Observe seed pods develop	17 18 19 20 21 22 23 24 25 26 27 28 29 30 31 32 33 34 35
13	Make model Brassica	22 23 24 25 26 27 28 29
	Check water level	26
14	Make model bee	29 30 31 32 33 34 35 36
15	Interpret graphs	36 37 38 39
	Remove plants from water	36
16	Harvest and thresh seeds	After 42
	Post-unit assessments	After 42

Key: ■ Do activity on this day   □ Do activity on one of these days

Teaching Strategies and Classroom Management Tips

The teaching strategies and classroom management tips in this section will help you give students the guidance they need to make the most of their hands-on experiences in this unit. These strategies and tips are based on the understanding that students already have knowledge and ideas about how the world works. And that useful learning results when they have the opportunity to think about their ideas as they engage in new experiences and encounter the ideas of others.

Classroom Discussion: Discussions effectively led by the teacher are important. Research shows that the way questions are asked as well as the time allowed for responses can contribute to the quality of the discussion. When you ask questions, think about what you want to achieve in the ensuing discussion. For example, open-ended questions, for which there is clearly no one right answer, will encourage students to give creative and thoughtful answers. Other types of questions can be used to encourage students to see specific relationships and contrasts or to help students to summarize and draw conclusions. It is good practice to mix these questions. It also is good practice to always give the students "wait-time" to answer; that time (some researchers recommend a minimum of 3 seconds) will buy you broader participation and more thoughtful answers.

Brainstorming: A brainstorming session is a whole-class exercise in which students contribute their thoughts about a particular idea or problem. It can be a stimulating and productive exercise when used to introduce a new science topic. As students learn the rules for brainstorming, they will become more and more adept in their participation.

To begin a session, define for students the topics about which ideas will be shared. Tell students the following rules:

- Accept all ideas without judgment.

- Don't criticize or make unnecessary comments about the contributions of others.

- Try to hitch your ideas onto the ideas of others.

Ways to Group Students: One of the best ways to teach hands-on science lessons is to arrange students in small groups of two to four. There are several advantages to this organization. It offers students a chance to learn from one another by sharing ideas, discoveries, and skills. Students also will develop important interpersonal skills that will serve them well in all aspects of life. Finally, by having students work in groups, you will have more time to work with those students who need the most help.

As students work, often it will be productive for them to talk about what they are doing, resulting in a steady hum of conversation. If you or others in the school are accustomed to a completely quiet room at all times, this new busy atmosphere may require some adjustment. It will be important, of course, to establish some limits to keep the noise under control.

Safety: This unit contains nothing of a highly toxic nature, but common sense dictates that nothing be put in the mouth. In fact, it is good practice to tell your students that, in science, materials are never tasted. Students may also need to be reminded that certain items, such as toothpicks, forceps, and water droppers, are not toys and should be used only as directed.

The anti-algae substance in the felt squares is copper sulfate, which dissolves in the water to prevent algae growth. As a solid, copper sulfate can be irritating to the skin; also, small amounts of the solid can be toxic and must not be ingested. However, the amount of copper sulfate used in the felt squares is very, very small and very dilute. Also, copper sulfate is not known to be an allergen. Finally, like all flowering plants, Wisconsin Fast Plants™ produce pollen, which can be an allergen.

Handling Materials: To help ensure an orderly progression through the unit, you will need to establish a system for storing and distributing materials. Being prepared is the key to success. Here are a few suggestions:

- Check your Timetable for Wisconsin Fast Plants daily. Know which activity is scheduled and what materials are going to be used.

- Familiarize yourself with the materials as soon as possible. It might even be useful to label everything and spread it out on a table for students to see.

- Organize your students so that they are involved in distributing and returning materials. If you have an existing network of cooperative groups, delegate the responsibility to one member of each group.

- Organize a distribution station and train your students to pick up and return supplies to that area. Depending on the activity, this might be as simple as picking up a seed and a hand lens, or as complicated as assembling all of the supplies needed to plant the seeds. A cafeteria-style approach works especially well when there are large numbers of items to distribute.

- Preview each lesson ahead of time. Some have specific suggestions for handling materials needed that day.

Additional management tips are provided througout the unit. Look for following icon.

Refer to the **Unit Overview** and **Materials List** for more materials information.

The Lighting System: The most important component of the life support system for Wisconsin Fast Plants™ is the lighting. These plants have been selectively bred to grow under continuous (24 hours a day) cool white fluorescent lights. Once the seeds have been planted, the lights must stay on. It would be wise to tape a note on your light bank saying "Please do not turn off the lights" and to inform the custodial staff who might routinely shut off lights left on in your building.

The growing tips of the plants should always be about 2 to 3 inches from the light bulbs. Naturally, as the plants grow, this will require that you adjust the distance by raising the light bank proportionally. Or, set the light bank at about 16 to 18 inches high, with the plants initially set at 14 to 16 inches high. Then lower the plants as they grow.

Setting Up a Learning Center: Supplemental science materials should have a permanent place in the room in a spot designated as a learning center. This learning center could be used by students in a number of ways: as an "on your own" project center, an observation post, a trade book reading nook, or simply as a place to spend unscheduled time when assignments are done.

In order to keep interest in the center high, it is a good idea to change it or add to it often. Here are a few suggestions of things to include:

- science trade books on plants, insects, or famous scientists

- snap-together centimeter cubes, a centimeter ruler, and a balance scale with an assortment of interesting objects to measure and weigh such as paper clips, a box of cereal, wooden blocks, a shoe, a lunch box, vegetables, canned goods, and stuffed animals

- magnifying glasses and an assortment of interesting objects to observe such as leaves, seeds, stems (celery is good), roots (like carrots), flowers (try dandelions or broccoli), insects, soil and rocks from the playground, newspaper (especially the comics), fabric scraps (burlap or anything coarse is best), sponges, chalk, salt, and feathers

- calculators

- audiovisual materials on related subjects, such as plants, insects, interdependence, pollination, life cycles, or famous scientists

- other live plants, especially those started from seed, that grow at a "normal" rate

- models of bees and *Brassica* plants

- items contributed by students for sharing, such as an insect collection, a honeycomb, magazine or newspaper articles featuring graphs, pictures, books, beeswax candles, pressed plants, seed collections, pods, and model or toy farm equipment

Curriculum Integration: There are many opportunities for curriculum integration in this unit. Look for the icons for math, reading, writing, art, and speaking that highlight these opportunities.

Evaluation: Evaluation suggestions are included at regular intervals throughout the unit. They can help you assess what students know and monitor how they are progressing. With that information, you can help provide targeted assistance to students who are struggling.

Appendix A offers post-unit assessments. A selection is offered so that the teacher can choose the most appropriate assessments for the students.

Appendix B provides a black line master for a "Teacher's Record Chart of Student Progress" that you may choose to use as a checklist. It could help you in your recordkeeping for the class and in your identification of students who are not keeping up. Please keep in mind that most third graders will not be able to master and articulate this full list of skills.

Portfolios of student work also are useful in many ways, for example, for sharing with parents or other interested adults. A portfolio could include student notebooks with drawings, writings, diagrams, completed activity sheets, and descriptions of projects. Finally, student presentations can be useful vehicles for assessment as well as useful language and critical thinking experiences in which students formulate and articulate their ideas.

What Do You Know about Plants?

Overview

In this first lesson, students will have an opportunity to reflect on how much they already know about plants and what they would like to learn. They are also asked to look closely at the outside of a bean seed, an exercise that introduces skills that will be developed throughout the unit: observing, recording, and predicting.

Objectives

- Students share what they know about plants and discuss what else they would like to know.

- The teacher evaluates students' prior knowledge of plants.

- Students practice observation and prediction skills.

Background

Observation is one of the main skills to be taught and developed in this unit. In some cases, scientists use only their five senses to study their subject. In other cases, they may use equipment, such as hand lenses or microscopes, to enhance their senses. In this lesson, students will be asked to use their eyes, nose, and fingers to observe dry bean seeds. (Tasting is not permitted in this unit.)

Predicting is another skill that is woven into the unit. A prediction is much more than a guess, because it is based on observations, experience, or scientific reasons. Students should be able to give reasons whenever they make predictions.

There are two ways to use the hand lens effectively. They are shown in Figure 1-1:

- Place the hand lens close to your eye, where a lens would be if you were wearing glasses. Hold the object in your other hand and move it back and forth slowly until it is in focus.

- Or, have students hold the object stationary while keeping the hand lens, or magnifier, above the object. Move the magnifier back and forth to focus.

Figure 1-1

Using a hand lens

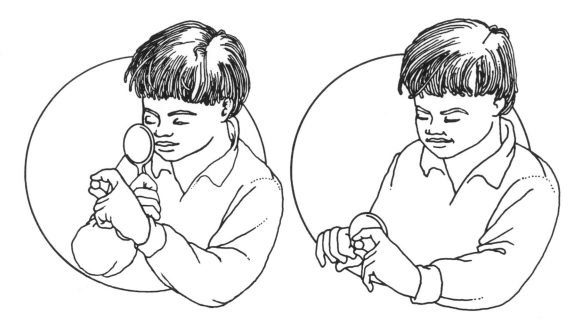

Materials

For each student
1   dry lima bean seed
1   student notebook
1   **Activity Sheet 1, Recording Chart for Seed Observations**

For every two students
1   hand lens

For the class
1   container of water
30-60   extra bean seeds

To record student ideas and questions, have one of the following available:
2   (or more) large sheets of newsprint and markers
    OR
2   (or more) overhead transparencies and markers with a projector

Preparation

(~15 minutes)

1.   Make copies of **Activity Sheet 1**.

2.   Get the materials needed to record student ideas and questions.

3.   Label one newsprint or overhead sheet "What We Know About Plants." Label the second sheet "What We Would Like to Know About Plants."

4.   Set up a distribution station for the seeds and hand lenses. In the first two lessons, students will practice picking up and returning the few supplies they are using at that time. The experience will come in handy in Lesson 3, when students will be asked to manage a much larger number of materials.

5.   Fill the container with water. Have students drop their seeds into the container at the end of the lesson.

Procedure

1. Tell your students that they will need a notebook for this unit. Since there will be a number of activity sheets accompanying the lessons, a looseleaf binder or a notebook with pockets will be the most efficient. Tell students that they will use their notebooks to record the life cycle of their plants. Later, they will be able to look back at their notebooks to remember some of their experiences and to see what they have learned.

2. Then introduce the unit by explaining to the class that for the next six weeks they will be observing the growth and development of a special plant. However, before they begin, you'd like them to share what they already know about plants and what questions they would like to have answered.

3. Using brainstorming techniques, ask students what they already know about plants. Remind them that during brainstorming sessions everyone shares ideas freely. Tell students that often someone else's ideas make us think of new things or make new connections. Stress the importance of accepting all ideas and of not criticizing anyone for making a contribution.

4. Display the sheet entitled "What We Know About Plants." Use it to record all student responses as objectively as possible.

5. Display the second sheet, "What We Would Like to Know About Plants." Record student questions.

   Keep both sheets. Use them as a:

   ■ **Pre-Unit Assessment**: Students have shared important information about what they know about plants. Build upon their level of knowledge and experiences. As you teach each lesson, have students add new ideas to the list. Write the new information with a different colored pen.

   ■ **Reference Point**: Refer to the comments and the questions raised as you teach the unit. By the end of the unit, many of the questions will be answered.

   ■ **Post-Unit Assessment:** Display the lists after you have completed the unit. Let students review the lists to evaluate their own progress.

6. Then, for the second part of the lesson, tell students that they will have an opportunity to look at a bean seed closely. Show students where the containers of bean seeds and hands lenses are laid out. Explain that you expect them to pick up and return their own supplies.

7. Ask students to look at the bean seed with the hand lens. Circulate to check that everyone is using the hand lens properly.

   Note: Students using magnifiers tend to forget that they have other senses that can also give them useful information. Encourage students to feel the seed and smell it. Do not allow any tasting, however. It is never a good idea to taste anything during science experiments.

8. Distribute **Activity Sheet 1**. Explain that it is important to keep a record of the seed observations. Tell them that this is especially true because the final part of the lesson will be to soak the bean seeds in water. Therefore, it is important to have a record of what the seeds were like when they were dry.

9. Preview the activity sheet with the class. Help students identify the different senses that contribute to making observations. Then ask them to complete the chart by writing one or two descriptive words in each of

What Do You Know about Plants? / **11**

the first four blocks (Color, Shape, Texture, and Odor) under the heading, "Dry Bean." Point out that they need to make a drawing in the last block, "Size." Students either may trace around the dry bean or lay it on a centimeter ruler to measure it. The importance of this measurement will not become obvious until the next lesson, when they observe that the soaked bean has taken in water and swelled.

10. Have students clean up by putting the hand lenses back in the container. Have students put the recording charts in their notebooks. Finally, ask students to put their dry bean seeds into the container of water you have set out.

Final Activities

Tell the class that their beans will sit in water overnight. Ask them to predict what they think will happen to the seeds, and to give reasons for their predictions. Record their predictions on a large sheet of newsprint.

Extensions

1. Ask students to do a seed survey at home. How many different kinds of seeds or seed products can they find? Remember that besides the obvious ones such as oats, peas, and beans, there are many less obvious seed products, such as coffee, peanut butter, corn chips, mustard, sunflower seed oil, and chocolate.

2. By this time, you probably have set up a learning center for students. (See **Tips for Classroom Management** for information about setting up a learning center.) Students may want to bring in examples of seeds to add to the center. There are whole seeds hidden in their lunch foods (apples, tomatoes, oranges, and grapes, for example), seeds out on the playground (acorns, dandelions, and pinecones), and some at home.

3. Consider an art project using seeds, such as making a seed mosaic or decorating a small can with seeds to be used as a pencil holder. (Frozen juice cans work well.)

Evaluation

Use these early assessments to measure students' progress as they work on the unit. Make note of the following:

Students' **previous knowledge** of plants:

■ How much information do students already have?

■ Have the students had previous experience planting, transplanting, pollinating, harvesting, or threshing?

Students' **observational** skills:

■ Are students able to make observations using all the senses (except taste)?

■ Can students distinguish between observations ("It's green.") and subjective statements ("It's gorgeous.")?

Students' **communication** skills:

■ Can students discuss seeds using observable properties?

■ Can students organize their observations into a chart?

Add 30 to 60 extra bean seeds to the container of water that has bean seeds soaking overnight. Check to make sure that water is covering the seeds completely before leaving. The next day, rinse the seeds, and then make sure that each child has 2 to 3 seeds to examine. If you leave the seeds soaking for longer than 24 hours, they will begin to ferment and smell bad.

Recording Chart for Seed Observations   **Activity Sheet 1**

NAME: _____

DATE: _____

Directions: In the spaces below, write one or two accurate words to describe the seed you are observing today.

	Dry Bean	Soaked Bean
Color		
Shape		
Texture		
Odor		
Size		

In this space, **draw** the dry bean seed

In this space, **draw** and **label** the inside of the soaked bean seed

What Is Inside a Seed?

Overview

In Lesson 1, students looked at the outside of a bean seed. In this lesson, students will look inside to observe the internal structures. *Brassica* seeds are too small for this experiment; bean seeds are bigger, which allows students to see the characteristics that most seeds share.

Objectives

- Students observe how the bean seed has changed after being soaked in water overnight.

- Students record their observations.

- Students open the bean and observe the inside.

- Students draw and label the parts of a bean seed.

Background

All seeds have two main parts: the **embryo**, or baby plant, and the cotyledon, a thickened leaf that stores food. The **embryo** is the tiny, undeveloped plant that includes all the parts of a mature plant. The outside of the seed, or the seed coat, protects these delicate internal structures. The cotyledon contains food—oils, carbohydrates, and protein. This provides energy for the initial underground growth of the plant. Once the plant has emerged and developed leaves, it can manufacture its own food. Figure 2-1 shows the inside of a bean seed.

Figure 2-1

Inside a bean seed

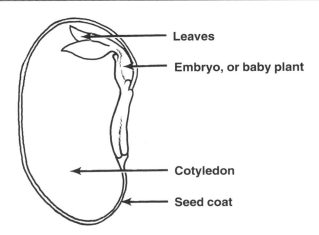

Materials

For each student

    1   **Activity Sheet 1, Recording Chart for Seed Observations**
    1   dry bean seed
2 or 3   soaked bean seeds
    1   paper towel
    1   student notebook

For every two students

    1   hand lens

For the class

    1   overhead transparency (Figure 2-2) with projector
        and screen (optional)

Preparation
(~15 minutes)

1. Set up the overhead projector and screen if you plan to include the optional activity. (See **Extensions** on pg. 17.) Reproduce the transparency.

2. Place the supplies in a central location.

Procedure

1. During the first part of the lesson, briefly review the five senses with the class. Discuss the kind of information each of the senses communicated about the dry bean seeds. This subject is probably very familiar to students. Discussing it will warm up the class and build confidence.

2. Bring out the list of student predictions about what they thought would happen to bean seeds soaked overnight. Discuss their ideas briefly. Then, tell them that today they will use as many of their senses as possible (except taste) to observe the soaked seeds. They will also have a dry seed so that they can compare the two.

3. Have students pick up their supplies and begin their observations of the seeds. Encourage students to use as many senses as possible to gather information about the outside of the soaked seeds. Some students may describe the seed coat as wet, soft, or wrinkled. Others may notice that the seeds now have a faint odor.

4. Tell students to record their new observations of the outside of the soaked seed on **Activity Sheet 1**, next to yesterday's observations about the dry seed. Ask them to notice the differences.

5. Next, tell students to peel off the seed coat. Then have them gently pry open the two halves of the seed. It is easiest to work from the outside of the curve on the side opposite the seed scar, the place where the seed was attached to the pod of the parent plant.

6. Circulate around the class to make sure all students have located the two main structures (embryo and cotyledon) and are using the hand lenses correctly. Encourage students to use the correct terms as they point out the parts of a seed.

7. Pass out the extra seeds to give students a second chance to dissect. Just as real scientists do, students have a chance to see if they can replicate their findings.

8. Finally, direct students to draw what they saw on their observation sheets. Stress that the drawing need not be a work of art, but should be a clear, complete, and accurate record of the information they learned about the seeds through their senses.

9. Direct students to clean up. Have them put their activity sheets in their notebooks and throw away their seeds and towels. Tell students to put hand lenses back in their designated container.

Final Activities

1. Ask students to share their observations of the seeds. Have them identify which sense they used to make a particular observation. Remind students that not all the senses are used for every observation.

2. Ask students how the seed changed after being soaked. How close were their predictions to what actually happened? This would be a good time to point out that predictions are never really right or wrong. They are our thoughts and ideas about what we think will happen in the future. Usually, the more observations and experiences we have to back up our predictions, the more accurate they are.

3. Finally, ask students what they think will happen to the *Brassica* seeds after they are put into the soil and watered.

Extensions

Project the overhead transparency (Figure 2-2). Explain that it is an example of a scientific drawing. Ask students to point out features that make it a good, clear, complete, and accurate picture of a plant.

Evaluation

The student record-keeping segments will be valuable tools to assess student progress. Criteria you may want to use to evaluate drawings and observations are listed below.

1. *Clarity:* Above all, the descriptions need to be clear and precise.

2. *Completeness:* All parts of the seed must be observed and described.

3. *Accuracy:* What you see on paper should be as close as possible to the real thing.

4. *Appropriate use of vocabulary:* Students should be able to incorporate the newly acquired words into their written observations. Also, they should be able to label the plant parts correctly.

In the next lesson your class will be planting the seeds. Try to recruit two or three adults to help.

Figure 2-2

Scientific drawing

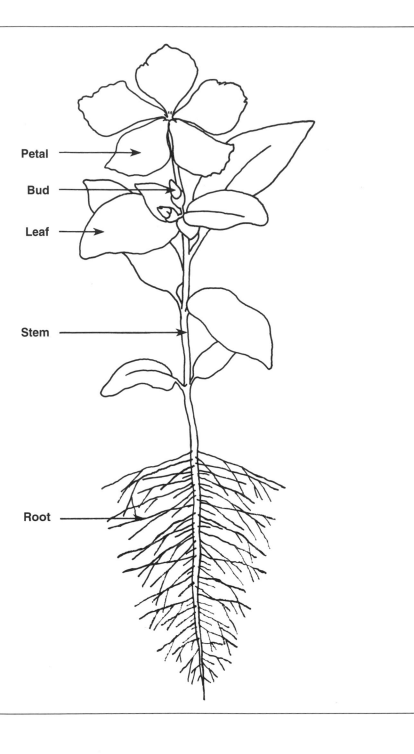

Petal

Bud

Leaf

Stem

Root

| LESSON 3 | # Planting the Seeds |

Overview

During this lesson, students plant the seeds. This activity is the basis of just about everything that follows, so it is extremely important that it be done correctly. The *Brassica* plant's unique characteristics require special planting materials and methods. The materials and methods needed are explained in detail in the lesson. There also are suggestions of how to prepare the students for planting day by introducing them to the supplies and by using **Activity Sheet 2** as a reading and sequencing activity.

For the best results, seeds should be planted on a Monday or Tuesday.

Objectives

- Students collect and organize their own materials for planting.

- Students set up their planters with wicks, fertilizer, potting mix, and seeds.

Background

The success of this lesson depends on careful organization and preparation. There are three important setups that you must do before students can begin planting: constructing the lighting system, arranging the watering system, and preparing the materials for distribution. For instructions on setting up the lighting system, see the insert enclosed in the lighting system kit; the other two procedures are explained in the **Preparation** section of this lesson. Read the instructions carefully.

Because this is a long lesson, you might want to spread it out over two days. Students can prepare the planters on the first day (Steps 1 through 5 on **Activity Sheet 2**) and complete planting the seeds on the next day (Steps 6 through 11 on **Activity Sheet 2**).

Materials

For each student

1   **Activity Sheet 2, How to Plant Wisconsin Fast Plants™ Seeds: Instructions and Checklist**
1   tray for supplies
1   planter quad

1 3-ounce cup of potting mix
12 fertilizer pellets
1 toothpick
1 paper towel
1 spoon
4 wicks
8 Wisconsin Fast Plants™ seeds
1 planter label

For every two students
1 pair of forceps
1 cup of water and dropper

For the class
6 to 8 sponges
2 dishpans of water
1 plastic-lined trash can
1 dustpan and whisk broom

Preparation
(~1 hour)

Below is an explanation of how to set up the watering system and tips on how to organize distribution of materials, student work spaces, and cleanup.

The Watering System

The unique watering system delivers water from the tank by capillary action. The route the water follows is from the tank through the mat and the wick to the potting mix. Wicks are inserted into each planter quad. The wicks come in contact with the mat, which hangs down into the tank of water. Below are instructions for setting up the watering system. Figure 3-1 illustrates how the watering system works. Once the system is set up, all you need to do is refill the tank every four to five days.

- Fill the water tank to capacity.

- Place the water mat in the tank to soak for about 15 minutes. Squeeze the mat out once or twice during this time and replace it in the tank so that it gets thoroughly saturated.

- Without squeezing the mat out, lay it on top of the tank's lid with the end dangling in the water. Smooth out any air pockets.

- Drop the blue copper sulfate squares into the tank to prevent algae from growing and clogging the system.

- After planting, double-check to see that each quad rests completely on the watering mat. Check the water mat and the potting mix in the quads each day to ensure that both are wet. This indicates that the wicking system is working. Continue to water from the top if the potting mix appears dry.

Three areas of the room need to be prepared in advance: the distribution station, student work spaces, and the cleanup area. Figure 3-2 shows how to arrange each of these areas.

Figure 3-1

Watering system

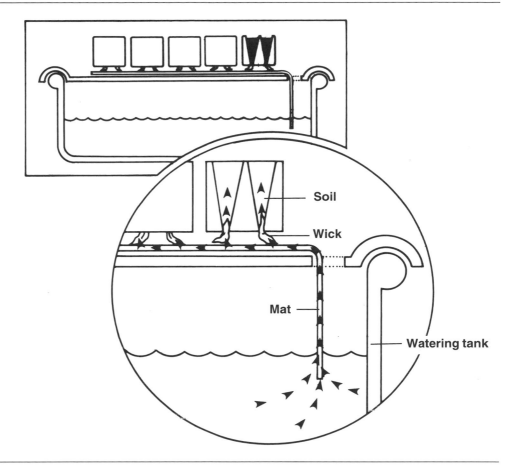

The Distribution Station

The distribution station includes all the materials needed for planting. Arrange the materials "cafeteria style," and have students pick up each item they need. This method has proven to be a real time-saver for teachers and a learning experience for students. To set up the distribution center efficiently, follow the guidelines listed below:

- Select one large area or several small areas of the room where students can easily walk by in single file on both sides of the supplies.

- Position all the materials in a line, on a series of desks or tables pushed together, or on the floor, if necessary.

- Ward off bottlenecks in the line by counting out the twelve fertilizer pellets and the eight seeds ahead of time and placing them in small cups.

- Moisten the potting mix if it has become powdery.

- Empty the potting mix into a large container. Have students dip their own cups into this container.

- Place a printed label on each item telling students what it is and how many to take.

Figure 3-2

Distribution station

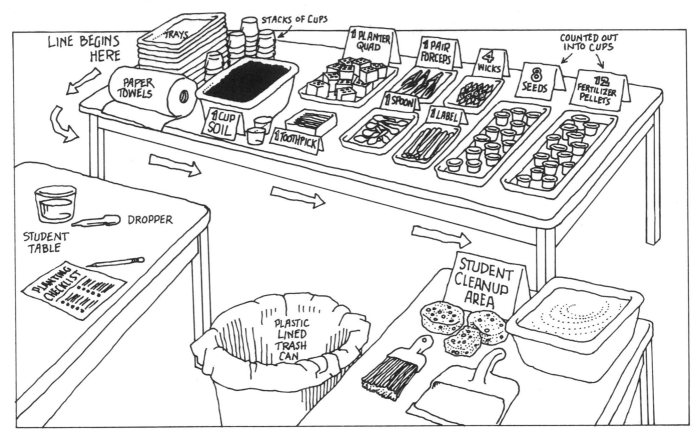

A SAMPLE SETUP for PLANTING DAY

Student Work Spaces

To help students work more efficiently, organize them as follows:

- Group students four to six to a space to make it easier for them to share watering supplies, forceps, and ideas.

- Clear the work area of any objects that might get wet or dirty.

- Have a copy for each student of **Activity Sheet 2**, the instructions and checklist for planting, at each work space.

The Cleanup Area

Supplies suggested for the cleanup area are a pan of water, sponges, whisk broom and dustpan, and a plastic-lined trash can.

Throughout the unit, students will be expected to clean up independently, thoroughly, and cheerfully. Make sure you place the cleanup area in a prominent and easily accessible location.

Be sure to plant extra quads of plants. You might need them for students who are absent today or to replace plants if needed later in the unit.

Procedure

1. Distribute **Activity Sheet 2**, the planting instructions and checklist. Preview it with the class, perhaps even the day before. To make sure students understand the instructions, assign them as a reading exercise. Then cut the numbers off, cut the list apart, and use the items as a sequencing activity.

2. Now, focus student attention on the distribution station. Demonstrate how to walk carefully, take turns, and read the labels. Arrange students in groups of four to six at work spaces. Then have each student pick up a set of supplies. Students will have to share some items (hand lenses, forceps, water and dropper), but they will be planting the seeds independently. Figure 3-3 shows the size difference between the seeds and fertilizer pellets.

Figure 3-3

Fertilizer pellets and seeds

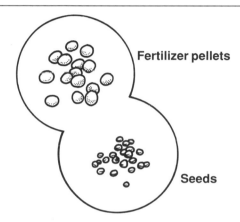

3. After students have returned to their work spaces, circulate around the room, assisting where necessary and assessing where appropriate. **Activity Sheet 2** is a good tool to assess how well students can follow directions. Which students are having problems, and where? Which students can work independently?

4. As work progresses, double-check some of these trouble spots:

   ■ Be sure the fertilizer goes in first, not the seeds. To point out the difference in size, you might lay both of them on the overhead projector.

   ■ Seeds planted too deeply will not germinate, or sprout, properly.

   ■ A great splash of water will wash out the seed. Each section must be saturated gently and thoroughly for wicking to begin.

   ■ Check that all quads are sitting under the light bank on the water mat before leaving for the day.

5. Cleanup is extensive and requires that everyone help. The more specific your instructions, the better students will do. Decide ahead of time exactly how you want the room to look at the end of the hour, and let the class know that nothing less will do. Encourage everyone to do a fair share of the work.

Final Activities

1. Remind students of what they learned about seeds in Lessons 1 and 2. Point out that they have taken dry seeds, placed them in potting mix, and added water. Ask them to predict what might happen to the seeds over the next 24 hours. Have them write their prediction in their notebooks.

2. Have students read *Fast Plants for Fast Times* on pg. 1 of the Student Activity Book (pg. 25 in the Teacher's Guide).

Extensions

Allow students to plant a variety of "normal" seeds to serve as comparisons to Wisconsin Fast Plants™. Some seeds to try include leftover beans from Lessons 1 and 2 or seeds students contributed to the learning center.

Evaluation

Evaluate student performance based on the following observations:

1. The ability to follow a written set of instructions to perform a task.

2. The degree to which a student is able to complete a project with no help from the teacher.

3. The student's ability to make a reasonable prediction of what will happen to the seeds based on previous observations and experiences.

For the next three days (or until you are sure that the wicking system is working), the planters must be watered from the top. If potting mix in the planters appears dark and feels moist when you arrive in the morning, you can be sure that the wicking has begun. After that, all you need to do is keep the water tanks filled.

**Reading
Selection**

Fast Plants for Fast Times

The Wisconsin Fast Plant™ is the plant you will be using for your experiments in this unit. It took Dr. Paul Williams, who is a professor and researcher at the University of Wisconsin, about fifteen years to develop it. Fifteen years may seem like a very long time to spend breeding a plant, but think of all that he accomplished. Through selective breeding, Dr. Williams was able to speed up the plant's life cycle, making it ten times faster than that of its ancestors. Today, this small, yellow-flowered plant whizzes through its entire life cycle, from seed to seed, in just 6 weeks,

Dr. Williams had an interesting reason for wanting to develop a fast plant. He is a plant pathologist, and his job is to study plant diseases and to find out if some plants inherit the ability to fight off diseases. In order to speed up his work, he needed a fast-growing plant to use in his studies.

Dr. Williams started with a world collection of more than 2,000 *Brassica* seeds and planted them in his laboratory using planting, lighting, and watering equipment almost exactly like what you will use. He observed that out of the 2,000, only a few plants flowered much sooner than others. He took advantage of these exceptional plants by cross-breeding them. These few would be the parents of his next generation of plants. Dr. Williams wondered what kind of offspring these faster flowering parents would produce. Would the offspring inherit the ability to flower earlier than the average *Brassica* plant?

Yes! In fact, a few of the new plants even flowered a little faster than the parent plants. These slightly faster offspring were then cross-pollinated, becoming the parents of the next generation.

Dr. Williams continued to use this method of selective breeding for years. He grew populations of 288 or more plants in each generation. He cross-bred the earliest flowering plants of this population and used their seeds to grow the next generation. In each new generation, he found that about 10 percent of the plants flowered slightly earlier than their parent generation had.

The selective breeding project was a grand success. The result is what is now known as Wisconsin Fast Plants™. Besides developing a 6-week growth cycle, Dr. Williams was able to breed in other desirable qualities that make the plant a nearly ideal laboratory tool. Some outstanding traits of these plants are:

- They produce lots of pollen and eggs, resulting in many fertile seeds.
- Their seeds do not need a dormancy (or rest) period, so they can be replanted immediately.
- The plants are small and compact.
- They thrive in a crowd.
- They grow well under constant light.

Wisconsin Fast Plants™ have become important laboratory research tools all over the world. Soon they will be part of National Aeronautics and Space Administration's space biology program. But most exciting of all, these special plants are becoming part of school science programs across the country, from the elementary to the university level.

How to Plant Wisconsin Fast Plants™ Seeds: **Activity Sheet 2**
Instructions and Checklist

NAME: _____

DATE: _____

Wisconsin Fast Plants™ are special in many ways. You must follow special directions when planting the seeds. It is very important to follow the directions carefully. Do one step at a time. Check off each step when you finish it.

☐ 1.   Pick up all of your **supplies** from the distribution station. Be sure you have these items before you begin planting:

___   1   planter quad
___   1   spoon
___   1   cup of potting mix
___   4   wicks
___   12 fertilizer pellets
___   8   Wisconsin Fast Plants™ seeds
___   1   toothpick
___   1   planter label
___   1   pair of forceps
___   1   paper towel

☐ 2.   Place one **wick** in each section of the planter quad. Use your forceps to pull the wick through the hole until the tip sticks out about 1 centimeter.

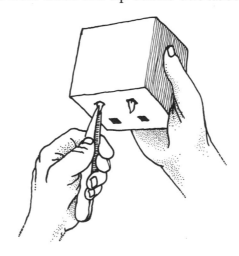

☐ 3.   Fill each section of the planter quad halfway with **potting mix**.

☐ 4.   Add **three fertilizer pellets** to each section. Look closely. The fertilizer pellets are much larger than the seeds.

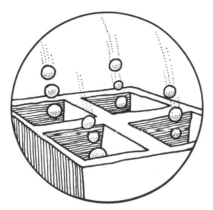

☐ 5.   Fill each section of the quad to the top with **potting mix**. Press it down a little with your fingers.

☐ 6. Put a drop of water on your tray and dip your toothpick in it. Use the wet toothpick to pick up one **seed**. Place the seed just below the potting mix and cover it. Plant a second seed in this section in the same way. Repeat until there are two seeds in each section of the planter.

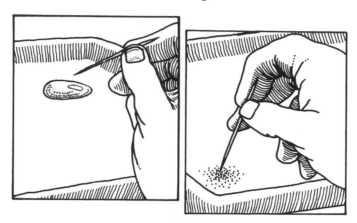

☐ 7. **Water** very gently, a drop or two at a time, until water drips from the bottom of each wick.

☐ 8. Write your name and today's date on the planter **label** and place it in the planter.

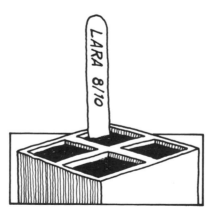

☐ 9. Place your quad under the light bank with the label facing out. Double-check to see that your planter is completely on the **water mat**. If you could see inside of each planter this is what it would look like.

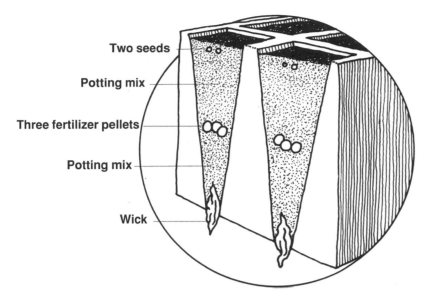

Two seeds

Potting mix

Three fertilizer pellets

Potting mix

Wick

☐ 10. Return all leftover **supplies** to the distribution station.

☐ 11. **Clean up your work space.**

Thinning and Transplanting (Day 4 or 5)

Overview

To maintain a healthy garden, people must periodically thin plants and transplant them to a more desirable location. As part of students' experience during this unit, they will be asked to thin to one plant per section and to transplant the surplus seedlings. This activity should be done on Day 4 or 5 after planting. A class discussion about these two practical gardening techniques helps students understand why they are so important.

Objectives

- Students discuss the purpose of thinning and transplanting.

- Students learn how to carry out these two tasks.

Background

Your students may balk at the idea of thinning. The excitement of planting and of watching the seedlings emerge is still so fresh. Who wants to pull out those tender young sprouts?

But thinning is important. As gardeners thin their beds to give their plants the best possible growing conditions, so must we. For the *Brassica* plants, it is best to thin to one plant per cell. This ensures that each plant will have ample space, light, food, water, and air circulation. Under these conditions, the plants will thrive and produce the highest possible seed yield at harvesttime. Plants grown under more crowded conditions must compete for these essentials and will be less productive.

There are several ways to thin. One way is to pinch off plants at the soil line. Another way is to cut them off at the soil line with scissors. A gentler way is to pull up the plant, root system and all, and transplant it to another location.

In this lesson, students will be given two choices of what to do with the plants they thin: pinching, or cutting off, or transplanting to cells where no seeds germinated.

Transplanting takes planning ahead and a gentle touch. Be sure each student has decided where the plant is to go before he or she uproots it. Minimizing time in transit for the plant is important to prevent root damage. The *Brassica* is a survivor. Most transplants probably will perk up a few hours after placement in their new cells although they may lag behind developmentally by a few days.

Materials

For every student

1   student notebook
1   toothpick
1   pair of scissors

For every two students

1   hand lens
1   pair of forceps (optional)

For the class

Potting mix
Surplus planter quads, if any
1   large or several small containers for the class plot (egg cartons,
    milk cartons cut lengthwise, margarine tubs)
Wicks for the above containers

Preparation

(~10 minutes)

1.   Distribute scissors and forceps (optional).

2.   If not enough wicks are left over, use cotton twine instead.

3.   Set up "class plots" with potting mix, wicks, and fertilizer. Place them in an easily accessible location so students can transplant their plants at their own rate.

Procedure

1.   Open the discussion by asking if anyone in the class has had experience thinning or transplanting. Follow up with a question asking why it is important to thin plants. Then discuss why sometimes it is necessary to transplant. Help students see that the purpose of both techniques is to improve growing conditions for the plant.

2.   Following the steps listed below and illustrated in Figure 4-1, have students spend the next 20 minutes observing and then thinning and transplanting.

   ■   After students have retrieved their plants, ask them to spend a few minutes observing the plants with a hand lens. Point out differences among the plants even at this very early stage of development. Ask: "Are all of your seedlings the same size? The same color? Where are the differences, exactly? In the shape or size of the leaf? In the length of the stem? Did every seed sprout, or germinate?"

   ■   Have students decide which one plant from each cell they will keep and which one they will thin out. They will end up with a total of four plants per planter, one per cell.

   ■   Before thinning, students should gently loosen the soil with a toothpick. Tell them to plan to set aside one of the extra seedlings so that they can draw it later.

   ■   Now students can thin their plants. They can either cut them off close to the soil and discard them or uproot them and transplant them.

3. Students have the following choices of what to do with the uprooted seedlings:

■ Transplant them into one of their own cells where no seeds germinate.

■ Donate them to a classmate for transplanting.

■ Transplant them into the prepared class pots.

Figure 4-1

Thinning and transplanting

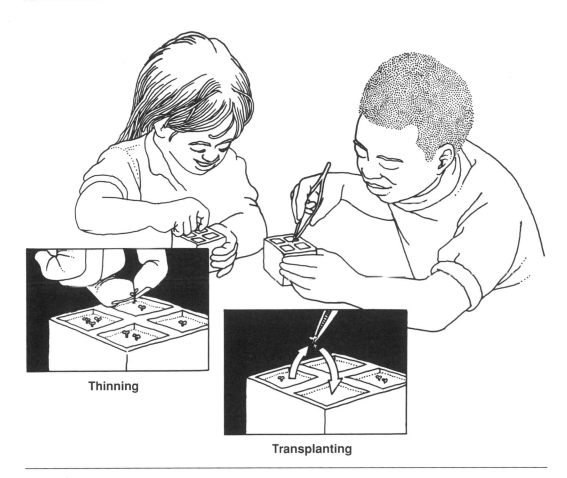

Thinning

Transplanting

4. After students have completed thinning and transplanting, have them draw and label one of their uprooted seedlings and write their observations in their notebooks. The drawing should include the seed leaves, the stem, and the roots. This is the only opportunity students will have to observe the roots. Remind students to include today's date and the age of the seedling with their observations.

5. Clean up. Have students throw away any plants you can't use and return equipment to their containers.

Final Activities

A closing discussion could focus on any of the following:

■ Observations of individual differences in seedlings that are exactly the same age. From there, point out that members of the class, though close in age, are different in important ways.

■ A review of the requirements for plant growth

- The optimum conditions for plant growth
- The differences between the bean seed embryo and the *Brassica* seedling

Extensions

1. Below are suggestions of how to use the transplanted plants:

   - Reserve some of the "rejects" to be used as replacements for plants that die unexpectedly from an accident or neglect.

   - Save the rest to use as experimental plants in the pollination lesson. For the time being, do nothing further with them except to see that their water wicking system is working. In about ten days, when the flowers open and students pollinate the plants, you can use these transplants as a control to see what happens when pollination does **not** take place. When it is appropriate, label these plants "Do Not Pollinate."

2. Encourage students to think about the questions listed in the Student Activity Book under **Ideas to Explore**.

Evaluation

To evaluate students during this lesson, consider the following:

1. Which students made contributions during discussions ?

2. How well do students work with their hands? Can they manipulate the tools easily? Do they handle the seedlings gently?

3. What was the quality of the notebook entry? Was the drawing complete, clear, and accurate? Were the words descriptive (green, heart-shaped leaves) rather than subjective (great leaves)?

LESSON 5	# How Does Your Plant Grow? (Day 5, 6, 7 or 8)

Overview

At this stage of the growing cycle (approximately Day 5, 6, or 7—possibly 8), the plants have grown tall enough for students to measure and record their height in centimeters on a graph. This is an opportunity for students to apply graphing skills to the experiences they are having growing plants.

Objectives

- Students learn how to measure their plants to the nearest centimeter.

- Students begin keeping records of their plant growth on a bar graph.

Background

Measuring, or making quantitative observations, is one of the most basic skills of science. One of the benefits of the exercise is that it stimulates students to make concrete comparisons precisely. Rather than saying "My plant is bigger than your plant," students can boast "My plant is two whole centimeters bigger than your plant!"

Of course, measurements may be done with a ruler, but students might enjoy the two other methods described in the **Procedure** section. Using paper strips to measure and graph offers a very concrete experience for the child. Using centimeter cubes, which are available in many classrooms, is more abstract.

To monitor the plants' progress accurately, measuring and graphing need to be done frequently throughout the unit. Because the plant will be growing most actively during the first 18 days of the cycle, it is *very important* to schedule the maximum number of measurements during that time. From Day 9 to Day 13, readings will be most dramatic because this is when the plants have their growth spurt. (Read more about this in Lesson 7.)

After Day 18, there will be relatively little upward growth. But it is important to continue measuring and graphing (perhaps only once a week) not only for additional practice but also to confirm that growth has indeed slowed. Students will discover that the reason for this is that the plant is busy developing flowers and seeds.

Materials

For each student

1 quad of plants
1 ruler marked in centimeters (optional)
1 sheet of centimeter graph paper (see **Appendix F**)
1 strip of paper precut to 1 centimeter wide (see **Appendix F**)
1 pair of scissors
1 student notebook

For the class

1 bag of 500 snap-together centimeter cubes
Glue
Crayons

Preparation
(~20 minutes)

1. Prepare the paper strips by holding the graph paper sideways and cutting it into strips 1 centimeter wide. (See **Appendix F** for a black line master of graph paper.)

2. Figure out how many centimeter cubes each student will need and pass out the appropriate number. As the plants grow, encourage students to estimate how many additional cubes they need and to borrow that number from a classmate.

Procedure

1. Distribute graph paper and other materials. Tell students to put the date and the age of the plant in the appropriate place. Tell students to select the one plant they will measure for the rest of the unit. They should move their name label to that section as a way to mark the plant chosen for measuring.

   Below are explanations of how to measure with paper strips and with centimeter cubes. Decide which method will work best for your class. Consider experimenting with both.

2. Demonstrate how to use the precut paper strip to measure the height of the plants. The steps are described below:

   ■ First, hold the paper behind the plant. The bottom of the paper should rest on an agreed upon spot (either on the potting mix or the pot rim, whichever you decide).

   ■ Mark the height of the plant by drawing a line on the strip.

   ■ Fill in the squares below that line.

   ■ Double-check by holding the paper up against the plant again.

   ■ Cut off the darkened squares.

   ■ Lay the darkened strip on the sheet of graph paper above the correct day number and paste it in place.

If you select the centimeter cube technique for measuring, pass out the cubes and give students a chance to manipulate them for a few minutes so that they become accustomed to how they fit together. (There is a high level of interest in the cubes. You may want to make them available to students at the learning center to use during unscheduled time.)

Figure 5-1

*Measuring with
paper strips*

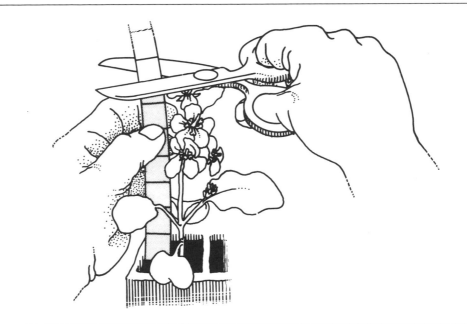

3. Demonstrate how to use the centimeter cubes as a measuring tool. Follow the directions below:

    ■ Tell students to fit the cubes together so they can measure their plants to the nearest centimeter. Remind students that the bottom cube must rest on the agreed upon place (either on the potting mix or the pot rim).

    ■ Lay the cubes on the graph above the correct day number and mark the height of the column.

    ■ Darken in the bar graph up to that height.

Figure 5-2

*Using cubes to
measure and graph*

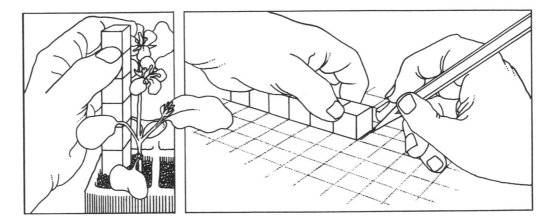

4. Instruct students to proceed with the activity independently, while you circulate, assist, and evaluate.

5. Tell students to give their graph a title and record today's date and the age of the plant. Have them place their graphs in their notebooks.

6. Clean up. Tell students to return all equipment to the distribution station and to throw away trash.

Final Activities

Ask students to explain the measuring and recording techniques they learned during this lesson. Tell them that you expect them to continue to use these techniques each time they measure and graph. Remind them that it is important to continue to measure the same plant each time. Have them double check that the date and age of the plant are recorded.

Extensions

1. Add some centimeter cubes, rulers, and paper strips to the learning center. Challenge students to measure an assortment of common objects in centimeters. Objects to consider are a paper clip, a coin, a bean seed, or an eraser.

2. Have students measure their foot length (from big toe to heel) and compare it with their forearm length (from wrist to elbow.) They will be surprised to learn that the two measurements are nearly identical.

3. Ask each student to cut and color a second strip representing the plant's height. Have one student collect these strips and sort them by height. Then set up a class graph showing number of plants and their height in centimeters on that day.

   Ask students to explain why all the plants are not exactly the same height even though they are the same age. Relate ideas about plants' heights to the range of students' heights in the classroom. Talk about normal variation.

Evaluation

Check for mastery of the following skills:

1. Students' manipulative skills:

   ■ Are there any problems cutting and pasting the paper strips?

   ■ Can everyone assemble the cubes into a column for measuring?

2. Students' precision in measuring and cutting.

3. Students' ability to interpret their own graph. Do they understand the purpose of the title? Can they read the two coordinates?

| LESSON 6 | # Observing: Leaves and Flower Buds (Day 7, 8, or 9) |

Overview

In addition to growing in height, plants also develop leaves and flower buds. In this lesson, students will observe the first true leaves and the buds on their *Brassica* plants and record their observations in both words and pictures. These developments will take place on approximately Day 7, 8, or 9.

Objectives

■ Students observe two major developments: the true leaves and the flower buds.

■ Students record their observations in their notebooks.

■ Students review the life cycle of a plant through this stage of development.

Background

In the life cycle of the *Brassica* plant, the embryonic root is the first part to emerge from the seed, usually within 24 hours of planting. About 24 hours later, the two cotyledons (now called the seed leaves) appear above ground, supported by the embryonic stem. The cotyledons, which were white while inside the seed, become green, and take on the job of producing food for the seedling until the true leaves develop. This takes place at about Day 7.

The seed leaves remain on the plant for some time after the true leaves have developed. Eventually most seed leaves will wither, die, and drop off. Right now, however, students can observe both kinds of leaves on the same plant. They will be able to see that the seed leaves are smooth and heart-shaped while the true leaves are larger, have wavy irregular edges, and a rougher texture.

Soon after the true leaves appear, the buds develop. They appear at the top of the stem in a cluster. The buds are closed tightly and are greenish-yellow in color. In another day or two, the buds will open to reveal the four-petaled, bright yellow *Brassica* flower.

This is a good opportunity to use the large Life Cycle Cards (**Appendix E**) to review the growth and development of the plant up to this point. Figure 6-1 shows the Life Cycle Cards and explains each stage of development.

Figure 6-1

Life Cycle Cards

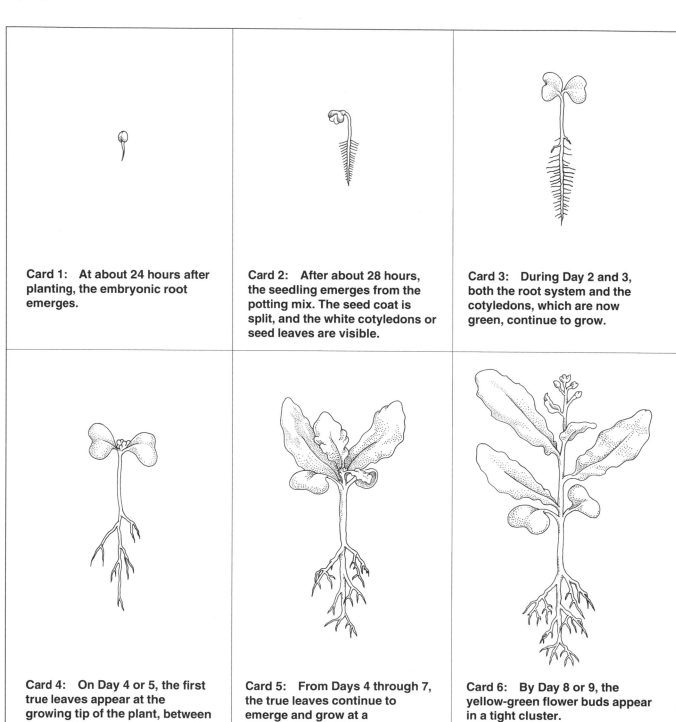

Card 1:   At about 24 hours after planting, the embryonic root emerges.

Card 2:   After about 28 hours, the seedling emerges from the potting mix. The seed coat is split, and the white cotyledons or seed leaves are visible.

Card 3:   During Day 2 and 3, both the root system and the cotyledons, which are now green, continue to grow.

Card 4:   On Day 4 or 5, the first true leaves appear at the growing tip of the plant, between the seed leaves.

Card 5:   From Days 4 through 7, the true leaves continue to emerge and grow at a remarkably rapid pace.

Card 6:   By Day 8 or 9, the yellow-green flower buds appear in a tight cluster.

Materials

For each student
1   quad of plants with buds
1   **Observation Sheet** (optional) (see pg. 43)
1   student notebook

For every two students
1   hand lens

For the class
Life Cycle Cards 1, 2, 3, 4, 5, 6 (**Appendix E**)

Preparation

Preview the information about Life Cycle Cards 1, 2, 3, 4, 5, 6 (see **Background**).

Procedure

1. Tell students to observe their quad of plants with a hand lens. They should notice the following:

    ■   Differences between the seed leaves and the true leaves

    ■   Number of both kinds of leaves

    ■   Color, size, shape, and number of buds

2. Encourage students to record their observations in as much detail as possible on their observation sheets. Suggest that they follow the steps outlined below:

    ■   First, tell students to draw and label the plant. Emphasize that the drawing should show the difference between the two kinds of leaves, should include the buds, and should have the right number of leaves and buds.

    ■   Next, have students write in their notebooks several descriptive sentences about the plant. Have them use the hints in the Student Activity Book (pg. 20) as guidelines.

    ■   Remind students to put the date on the observation sheet and to note the age of the plant in days. Have them place their **Observation Sheet** in their notebook.

3. To clean up, have students return plants to the water mat and put away the hand lenses.

Final Activities

1. Begin a review of the plant's life cycle up to this stage of development by asking students how old their plant is today. Remark that the plants have changed a lot since Day 1, and that this is a good time to talk about how the plant has grown and what new parts have developed. To begin the discussion, display the large Life Cycle Card 1, the seed.

2. Ask students to recount the plant's life cycle in the correct order. Continue to display the Life Cycle Cards in the correct sequence as the students describe the plant at each stage. Figure 6-1 shows the Life Cycle Cards and explains each stage of development.

3. Finally, ask students to predict what will happen next. Based on previous experience, many will know that the buds will open into flowers.

Extensions

1. Ask students to think of examples of leaves and buds that we eat. Students will probably know many edible leaves, such as lettuce, collard greens, spinach, watercress, parsley, and cabbage. But not many students will know that artichokes, broccoli, brussels sprouts, and cauliflower are edible buds and flowers.

2. Add leaves and buds to the science center for children to observe. In addition to the edible ones listed above, there are countless free and fascinating specimens outdoors. Depending on the season and the locale, you may be able to take a short field trip with the class to collect leaves or buds.

3. Use the leaf collection as the basis for an art project. Spatter painting and leaf rubbings are favorites.

4. Large buds such as magnolia, hollyhock, and brussels sprouts are excellent subjects for dissection. Using forceps and toothpicks, have students peel away the layers to see how complex the bud structure is and how efficiently it is packaged.

Evaluation

Frequent monitoring of student products will be a valuable evaluation and diagnostic tool. Look for continued progress in the following areas:

1. Are the students selecting observable properties to describe?

2. Are the observations clear, complete, and accurate?

3. Are students using the newly acquired plant vocabulary appropriately?

4. Do the drawings include labels, dates, and the age of the plant?

Observation Sheet

NAME: _____

Day #	Date	Illustration	Observations

Observing the Growth Spurt (Days 9 to 13)

Overview

Just as human beings have growth spurts, so do plants. For human beings, the growth spurt usually takes place during adolescence. For the *Brassica* plant, it usually happens some time between Day 9 and Day 13. During this time, students will closely monitor the growth of their plants by measuring their height, predicting overnight growth, and recording data. This lesson draws on measuring and graphing skills acquired in Lesson 5.

Objectives

■ Students measure plant height in centimeters and record it on a graph every day for one week.

■ Students predict how much their plant will grow each day.

■ Students analyze their data on the growth spurt.

Background

As mentioned above, the growth spurt usually takes place between Day 9 and 13. However, plants grow at vastly different rates. Student observations are scheduled to occur for one week between Day 7 and 14 to allow for early or late bloomers and to help establish the pattern of normal growth for the plants.

The students were introduced to two techniques for measuring and graphing in Lesson 5. After this initial lesson on the growth spurt, or even before, you may decide that students are able to observe, measure, predict, and graph independently.

Life Cycle Cards 6 and 7 should be displayed as a reminder to the students to continue with their observations. Card 6 shows the plant before the growth spurt with flower buds tightly closed. Card 7 shows the plant after the growth spurt with flowers fully opened. Figure 7-1 shows these two phases. It is interesting to notice that the upward growth takes place in the stem between the leaf nodes, places where the leaves are attached.

Materials

For each student
   1   quad of plants
   1   **Activity Sheet 3, Observing the Growth Spurt**
   1   sheet of centimeter graph paper (see **Appendix F**)

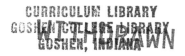

Figure 7-1

Growth spurt

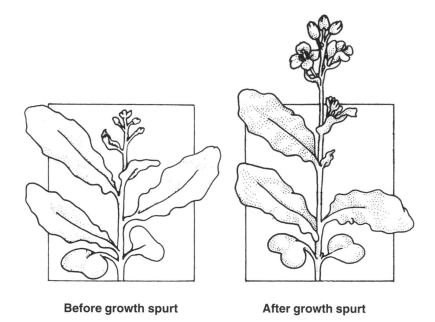

Before growth spurt   **After growth spurt**

1 strip of paper precut to 1 centimeter wide (see **Appendix F**)
1 pair of scissors
1 student notebook

For the class
   Glue
   Life Cycle Cards 6 and 7

Preparation

Make copies of **Activity Sheet 3** and of the graph paper for each student.

Procedure

1. Begin the lesson by telling students that today they will begin keeping records of their plant's growth spurt. This will involve making observations, taking measurements, and making predictions every school day for a week.

2. Review the correct ways of measuring the plant using either the paper strips or the cubes (see Lesson 5).

3. Preview **Activity Sheet 3** with the class. Before they begin filling out the chart, check that students understand the following:

    ■ what is meant by a growth spurt

    ■ how to record their data for the week

    ■ what they need to know to make a prediction

4. Allow students to proceed at their own pace in measuring and recording data. If students are to work independently at unscheduled times, establish the routine during this lesson. For example, make supplies available in an easily accessible location, discuss what times during the day it would be appropriate for students to carry out the activity, specify group size, and explain expected behavior.

5. Remind students to record the plant's height on their graph each day. Also, remind them to keep their sheets in their notebooks.

6. Check individual work as often as possible during the week.

Final Activities

After Day 15, it will be interesting to pool the class information about the growth spurt. Here are some discussion questions:

- When did your plant go through its growth spurt? What proof do you have? Give evidence to back up your claim.

- How tall was your plant before the growth spurt? How tall was it after the growth spurt?

- Look at your graph. What was the most your plant grew in 24 hours?

- How close were your predictions to what really happened? Did you make better predictions after you had some practice? Or was your plant unpredictable?

Extensions

1. Many people enjoy sharing tales of the year Mom had to let the hems down twice because of an adolescent growth spurt. Ask students to interview a parent or relative about their growth spurt.

2. Practice making predictions. Pick a topic of interest, make observations, collect data, and record the class prediction. Some good topics might be related to sports events, weather, or when a "first" of something in nature (such as a robin sighting or a snow fall) will occur for that season.

Evaluation

The students' products will serve as evaluation of:

- their ability to follow directions and work independently

- organizational skills

- record-keeping skills

- interest and persistence

- the ability to take accurate measurements

- the ability to make reasonable predictions based on evidence

Observing the Growth Spurt

Activity Sheet 3

NAME: _____

Day	Today's Date	Observations	Height Today	Height Prediction
			cm	cm
			cm	cm
			cm	cm
			cm	cm
			cm	cm
			cm	cm
			cm	cm
			cm	cm

Why Are Bees Important?

Overview

The first seven lessons introduced students to plants and how they grow. In the next several lessons, students will learn about pollination, the process through which plants are fertilized. For many plants, including the *Brassica*, the bee facilitates this process. In this lesson, students will have an opportunity to share what they already know about the bee and about pollination.

Objectives

■ Students share information about bees and raise questions about them.

■ Students draw a picture of what they think a bee looks like.

Background

The class has already had experience brainstorming in Lesson 1. Review with the class the guidelines for participating in a brainstorming session. They are listed below:

■ Accept all ideas nonjudgmentally.

■ Do not criticize or discuss other people's ideas.

■ State all ideas in a positive way.

■ Encourage students to draw on the ideas of others in developing their own ideas.

Materials

For each student
  1 sheet of drawing paper
     Crayons (optional)

For the class
To record student ideas and questions, use one of the following:
  2 (or more) large sheets of newsprint and markers,
  2 (or more) clean overhead transparencies, markers,
       projector and screen

Preparation

(~15 minutes)

1. Obtain the materials you need to record student ideas and questions.

2. Label one sheet "What We Know about Bees." Label the second sheet "What We Would Like to Know about Bees."

3. Obtain art supplies for the drawing segment of the lesson.

Procedure

1. Introduce the lesson by displaying the two recording sheets. Tell students that there will be a brief brainstorming session to find out what they already know about bees and what they would like to find out.

2. If necessary, review the rules for brainstorming.

3. Use the two sheets to record all student responses. Keep both sheets and use them as a way to understand what students know about bees and what their attitudes about them are. Refer to these sheets as needed throughout the unit.

4. After the brainstorming session, distribute art supplies. Invite students to draw a picture of a bee. Tell them that although you expect them to do the best job they can, you do not expect them to be bee experts.

Final Activities

Collect the drawings and save them. Keep them out of sight until the post-unit assessment on bees. Tell the students that at the end of the unit they will do another bee drawing. Then they will hang the two side by side and assess for themselves how much they have learned about bees.

Extensions

1. Bees are fascinating creatures and the subject of many excellent trade books, some of which are listed in the **Bibliography**. You might challenge your students to read to find out the answers to these questions:

   ■ *What is a "honey stomach"?*

   (The "honey stomach," or crop, is a nondigesting storage sac, which the bee uses to collect and transport nectar back to the hive.)

   ■ *Why do bees dance?*

   (Worker bees communicate information about the location of a rich supply of food by doing a dance upon returning to the hive. If the flowers are close by, the bee dances in a circle. This is called a round dance. If the flowers are more than 100 yards from the hive, the bee does a figure-eight dance.)

   ■ *Find out more about African honeybees. Why are they called "killer bees"?*

   (The so-called "killer bees" are a strain of wild African bees descended from a handful of African queens released in Brazil in 1959. They have been working their way north over the past 30 years and are expected to invade Texas in the 1990s. The African bees' venom is no more lethal than any other. Their reputation comes from the fact that they are easily provoked to attack, and they defend their hives with a vengeance. Tales of African bees stinging people en masse or chasing an intruder for miles are greatly exaggerated.)

Evaluation
Consider the lesson a pre-assessment on the subject of bees. Make note of the following:

1. How much do students know about:

   ■ bee anatomy

   ■ pollination

   ■ interdependence of bees and flowering plants

2. What are the students' attitudes about bees? Do they feel that bees are friends or foes?

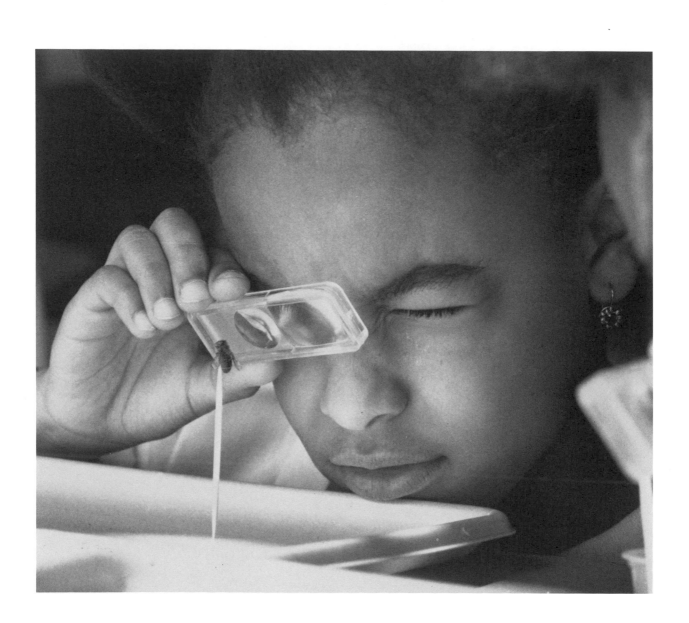

Getting a Handle on Your Bee

Overview

After speculating about bees in the previous lesson, students now have a chance to observe real bees that have been dried. These bees will be attached to sticks that can be used to actually pollinate the plants during the next lesson. Students enjoy learning more about bees and will be excited about the preparations to pollinate the plants.

Objectives

- Students use a hand lens to observe dried bees.
- Students make bee sticks to be used as a tool for pollination.

Background

Despite the many benefits we enjoy from the work of honeybees, many of us are afraid of them, or, at best, think of bees as pests to be warded off before they have a chance to sting. The bee doesn't deserve such bad press. It is a vital contributor to life on earth. In addition to producing wax and honey, the bee is a major agent of pollination, the process by which pollen is transferred from the male part of one plant to the female part of another plant. This allows fertilization and seed production to take place.

You undoubtedly will discover the full range of attitudes toward bees in your own class. At first, your students may express all kinds of negative reactions, and you yourself may share some of their sentiments. Don't be concerned. This is normal. Expect it.

As your students learn more about bees, a transformation will take place. Students will find it exciting to observe this otherworldly creature close up, especially since it is now harmless. Noise level will go up but if you listen carefully, conversation will probably be "bee" related.

After pollination has been introduced and students begin to use their bee sticks as tools, they will handle the creatures quite matter-of-factly. Many will even begin to take some pride in how well their bee sticks work. By the end of the unit, students will ask to keep their bees!

The Colony

The honeybee is a social animal that lives in a remarkably well-organized colony consisting of three kinds of bees: the queen, the drones, and the workers (the kind used in this unit). Each kind of bee has basically the same three-part body plan consisting of the head, the thorax, and the abdomen. But because each kind of bee has a different job to do, parts of their bodies have evolved in specialized ways. Below are illustrations and brief descriptions of the bees in the colony and the jobs they do (Figure 9-1). The illustration on pg. 57 shows a worker bee with its body parts labeled. The illustration is suitable for reproduction as an overhead transparency.

Figure 9-1

Three kinds of bees

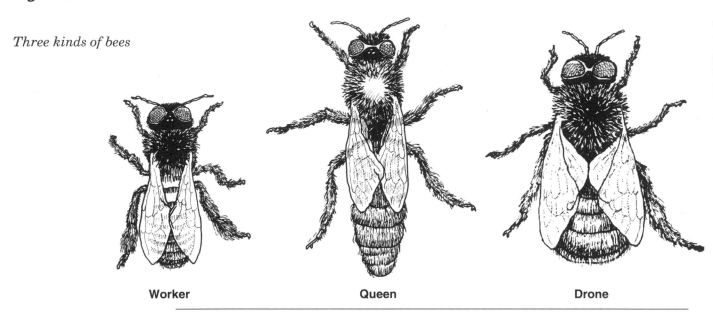

| Worker | Queen | Drone |

QUEEN   Life span: 3 to 5 years

The largest occupant of the hive and the only one of her kind, the queen bee is a virtual egg factory, capable of producing about 1,500 eggs each day. Shortly after hatching, the virgin queen takes one nuptial (mating) flight and is fertilized for life, with the drone's sperm stored in a special sac in her body. The queen then returns to the hive to begin laying eggs that will become either workers or drones.

DRONE   Life span: 1 or 2 seasons (spring or summer or both)

Stockier than the queen and a strong flier, the drone makes up about 10 percent of the hive population. His only purpose in life is to catch the queen during the nuptial flight and fertilize her. Ironically, the winner dies in the act. The rest of the drones return to the hive to be fed and cared for by the worker bees until food gets scarce in the fall. Then the workers bite off the drones' wings and unceremoniously throw them out into the cold.

WORKER   Life span: 3 to 6 weeks

Smallest in size but comprising 90 percent of the hive's population, this bee is always a sterile female. The hive could not exist without her, and she literally works herself to death. At different stages in her life she specializes in different tasks, such as feeding larvae; building, cleaning, and guarding the hive; secreting wax; controlling the hive temperature; and collecting nectar and pollen.

The body of the worker bee is specially adapted to collecting food. Her long, straw-like tongue siphons up nectar from deep inside the flowers. The nectar is then stored in a nondigesting honey stomach for transport back to the hive. The worker's hairy body traps pollen that the bee stores in a pollen basket on her hind leg. In this manner, both the nectar and pollen get carried back to the hive to feed the colony.

Materials

For each student
1 dried bee
1 toothpick
1 tray
1 pair of forceps
1 hand lens
1 **Activity Sheet 4, How to Make a Bee Stick**

For every four students
1 small cup of white glue
1 cup

For the class
1 overhead transparency, "The Worker Bee's Body" (see pg.57)
1 overhead projector and screen

Preparation
(~15 minutes)

1. Place the supplies listed above (except for the Activity Sheet) in the distribution station for each student to pick up. Group four to six students together so that they can share glue.

2. Place a small dollop of glue in a cup at each work table.

3. Place an inverted paper cup at each work table for students to poke their finished bee sticks into for storage.

4. Obtain an overhead projector and screen.

Procedure

1. Distribute **Activity Sheet 4** to the class and preview it together. The sheet provides all you and the students need to know about making bee sticks.

2. When everyone is clear on how to make the bee stick, direct children to the distribution station, where they will pick up their own supplies. It is best to delay this step until you are ready for students to actually begin working with the materials. Students will be interested in examining the bees, trying out the hand lenses, and manipulating the forceps. All of this is fine, and exactly what you want to happen, but not until instructions are clear.

3. Allow children time to examine their bees with the hand lenses. They may ask if the bee is real and if so, who killed it. Assure them that the bee is indeed real and died a natural death outside the hive at summer's end. The beekeepers collected all the dead bees and sold them to be used in this project.

4. Encourage children to work independently, following the instructions on **Activity Sheet 4**.

5. Circulate around the class to make informal assessments of student progress. **Activity Sheet 4** is a handy tool for assessing what has already been done and where help is needed.

6. When everyone has finished, collect all unused supplies and return them to the distribution station. Find a convenient place to store the bee sticks. Discard used glue cups and paper towels.

Final Activities

1. Project the overhead transparency "The Worker Bee's Body" (Figure 9-2) and allow students to observe it for a short time. (Perhaps while cleanup is going on).

2. Initiate an observation exercise by asking the children to describe the parts of the bee. Add that students may also tell what they think the part is used for.

3. Help students identify the parts of the bee's body. If appropriate, supply their names: head, thorax (or midsection), and abdomen. Also, point out additional features: the bee's two large eyes; its three small eyes, and the two antennae used for touching, tasting, hearing, and smelling.

4. Additional observations include noting the thorax and the six jointed legs and four wings attached to this section.

5. Finally, observe the abdomen, which is the bee's largest section and has the stinger at the end. It is also less hairy than the other two sections.

6. Ask students why they think they made the bee stick. What do they think they are going to do with it?

Extensions

1. Suggest to students who continue to show unusual interest in the bee that they do further reading in related trade books. See the **Bibliography** for book ideas.

2. Place several extra bees and some hand lenses at the learning center so children may observe at their leisure.

3. If someone has an insect collection, suggest that student share it with the class.

4. Tell students to be on the lookout for other insects to observe. Ask them to notice how other insects are like bees and how they are different.

Evaluation

This lesson reinforces skills taught in previous lessons. This lesson provides an opportunity to notice progress. Focus on these areas.

1. How detailed are the bee observations? Are students able to differentiate between observations and opinions?

2. To what degree are students able to follow instructions and complete the task independently?

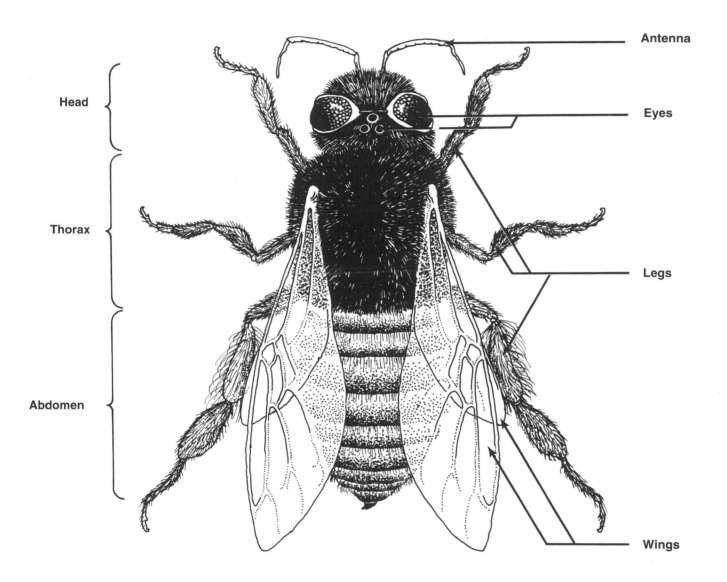

Head

Thorax

Abdomen

Antenna

Eyes

Legs

Wings

How to Make a Bee Stick   **Activity Sheet 4**

NAME: _____

DATE: _____

☐ 1.   Check off each item on the supply list to be sure you have everything you need before beginning.

     ___ 1 dried bee   ___ 1 cup of glue (for table)

     ___ 1 toothpick   ___ 1 hand lens

     ___ 1 pair of forceps   ___ 1 cup (for table)

☐ 2.   Observe the bee with the hand lens. Turn the bee over. Find the place where the legs are attached.

☐ 3.   Put a very small drop of glue on one end of the toothpick.

☐ 4.   Glue the underside of the bee (where the legs are) to the toothpick. Make sure the head is at the end.

☐ 5.   Let the glue dry for a few minutes. Be careful that the bee does not slip down the stick.

☐ 6.   Now, take the time to really observe the bee closely. How many body parts can you find? Check off each one.

    Note: Some bees may be damaged and not have all parts. Ask a classmate to share if your bee is not complete.

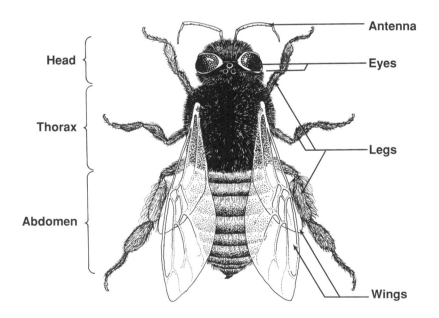

☐ 7.   Push the bee stick into the bottom of an upside-down paper cup. There is one cup for each table. Leave the cup at your table. Your teacher will store the cups.

☐ 8.   Place all supplies neatly on the tray in the center of your table.

Looking at Flowers

Overview

From Day 12 to Day 18, students will be using their bee sticks to cross-pollinate the flowers. This lesson will give students the opportunity to observe the flowers more carefully.

Objectives

■ Students observe details of the flower's anatomy and identify the major parts.

■ Students learn more about the crucifer family.

Background

The rapid-cycling *Brassica rapa,* which is the scientific name for Wisconsin Fast Plants™, belongs to a large family of plants called the Cruciferae. They are so named because the four flower petals are always arranged in the shape of a crucifix, or cross. Figure 10-1 (pg. 63) shows the *Brassica* flower. This same petal arrangement is shared by all members of this family of plants. A sampling of plants in the crucifer family is shown in Figure 10-2 on page 64.

Materials

For each student
    1   flowering plant
    1   student notebook
    1   **Observation Sheet** (optional)

For every two students
    1   hand lens

For the class
    1   overhead projector and transparency (for Figures 10-1 and 10-2)

Preparation
(~10 minutes)

1.   Set up projector and screen.

2.   Distribute equipment.

Procedure

1. Tell students to observe their plants with the hand lens. Have them pay special attention to the flowers. Allow sufficient time.

2. After all students have had a chance to observe the flowers, talk about the parts of a flower. Help students use their hand lenses to see four petals that are yellow and rounded; six anthers, two short and four tall; yellow pollen; and the pistil in the center. With the hand lens, they can see how the sticky stigma on top of the pistil glistens. Students are not expected to know the correct botanical names, however. Introduce the vocabulary as needed, and encourage the children to use it.

3. Have students record their observations of the flower in their notebooks. Ask them to draw the flower and to include all the parts. Provide the names and the correct spellings as needed for the labels.

   Figure 10-1 shows a *Brassica* flower with its labels.

4. Tell students to continue to observe the flower closely over the next week. They will notice the following dramatic changes:

   ■ Petals will fade and fall.

   ■ After pollination, the pistil will enlarge and become the seed pod.

   Note: You may not want to share this information with students right now. Instead, let them enjoy making the discoveries for themselves.

5. Have students read *The Crucifer Family* on pg. 31 in the Student Activity Book (pg. 61 in the Teacher's Guide).

Final Activities

1. Ask students to read the information about crucifers we eat on pg. 32 of the Student Activity Book (pg. 62 of the Teacher's Guide). Ask students which of these crucifers they have tasted. Use the graph paper in **Appendix E** to make a transparency. Record the results of the crucifer "taste" survey on the graph.

2. Review the parts of a crucifer plant with the class.

3. Remind students that it will be important to notice changes in the flowers over the next week.

Extensions

1. A florist may be willing to donate some "tired" blossoms to your class for dissection. A large flower (such as hollyhock, tulip, hibiscus, yucca, or lily) will have parts that are easy to see. Or, students can pick a flower growing in their neighborhood. Students can use toothpicks and forceps to tease apart the pieces. Have them tape all the parts to a piece of paper and identify the parts with labels.

2. Ask students to go on a crucifer hunt the next time they are in a grocery store. Ask them to pay special attention to those crucifers that were unfamiliar to them.

3. Bring in a stalk of old and yellowing broccoli for the learning center. Challenge students to find the flowers on this crucifer. Then work with the students and classify the crucifers listed on the crucifer survey according to what part we eat: root, stem, leaves, flowers.

4. Plan a lunch using plants in the *Brassica* family. Research recipes from faraway lands to try for your lunch.

Evaluation

Evaluate student progress in making observations. Check their drawings of the flower for accuracy and completeness.

Reading Selection

The Crucifer Family

It may seem odd to you that your plant belongs to a family, but it's true. Of course, it's not the kind of family with aunts and cousins and sisters. Think of it more as a group of plants that are alike in some ways.

Your *Brassica* plant belongs to the crucifer family. Crucifers have one thing in common. This feature gives the plants their family name. Here is a hint. The shape of their flowers is always like this:

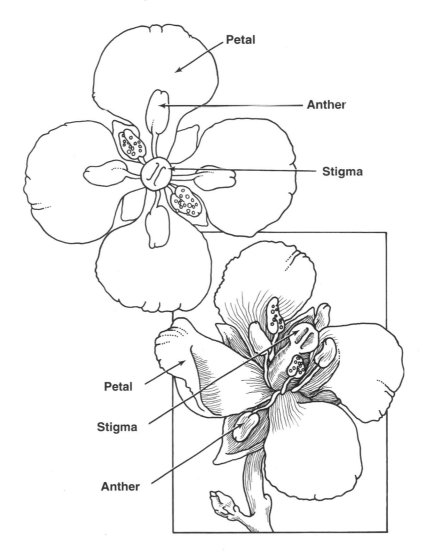

Did you figure out how the crucifer family got its name? The reason is that the crucifer flower always has four petals arranged in a cross. Scientists group all these flowers into the crucifer family.

A CRUCIFER SURVEY

The crucifers are an important food crop in many parts of the world. Which ones have you tasted?

Cabbage

Turnip

Collard

Watercress

Kohlrabi

Choy sum

Cauliflower

Broccoli

Rutabaga

Radish

Kale

Horseradish

Pak choi (Chinese mustard)

Brussels sprouts

Mustard greens

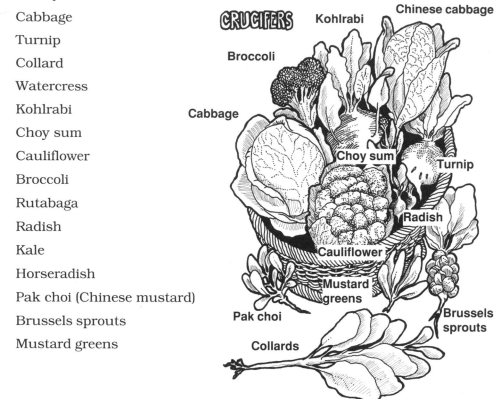

Some crucifer seeds are crushed for their oil. Others, like turnips, kale and rutabagas, are good food for sheep and cattle as well as for people. Still others, like alyssum and candy tuft, are known for their beautiful flowers. There is even a branch of "bad guys" in the family, some pesky weeds!

Figure 10-1

Brassica *flower*

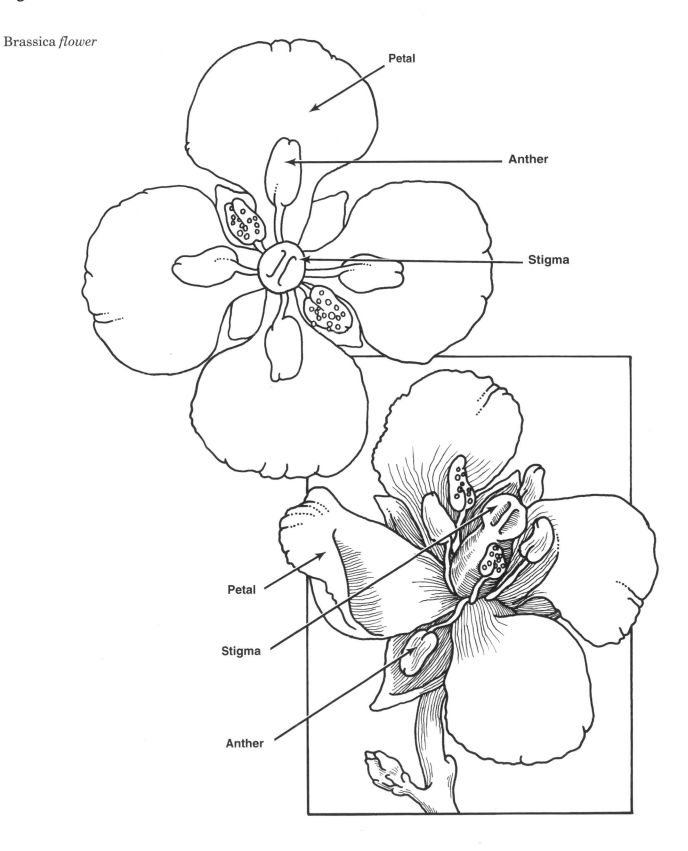

Petal

Anther

Stigma

Petal

Stigma

Anther

Figure 10-2

A crucifer survey

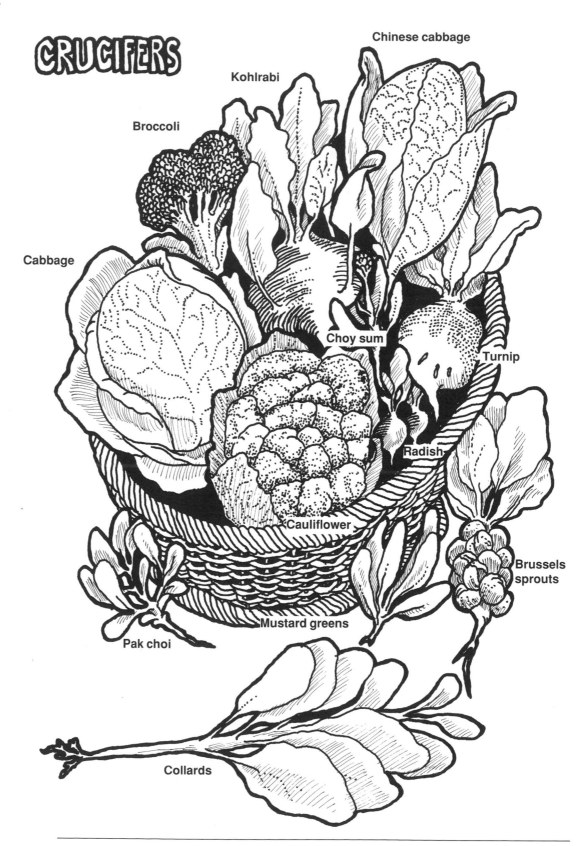

CRUCIFERS

Chinese cabbage

Kohlrabi

Broccoli

Cabbage

Choy sum

Turnip

Cauliflower

Radish

Brussels sprouts

Mustard greens

Pak choi

Collards

Pollinating Flowers
(Day 12 to 18)

Overview

For five consecutive school days (Day 12 through Day 18), students use their bee sticks to cross-pollinate their plants. Through discussions and readings, they will come to appreciate the interdependent relationship of the bee and the *Brassica*.

Objectives

■ Students use the bee sticks to cross-pollinate their plants.

■ Students read more about the interdependence of bees and flowers.

Background

There are many examples of symbiotic relationships in nature, where each partner is dependent on the other. For example, there is the mutually beneficial association of cattle and the cattle egret. The cattle are useful to the birds because they provide food—ticks that have a parasitic relationship with the cattle. The birds are useful to cattle because they free the cattle from these ticks. Between bees and flowering plants, the symbiotic relationship is miraculously complex.

To us, the flower is a thing of beauty, delighting our senses with color and perfume. But in nature, the flower is actually a specialized reproductive part whose sole purpose is to manufacture seeds. In the *Brassica* flower, the male parts (the six pollen-producing anthers) encircle the female part (the ovule-laden pistil). For an ovule to become a seed, it must be fertilized by a pollen grain from the same kind of flower. For the Wisconsin Fast Plants™, the pollen can **not** come from the same flower or from a flower on the same plant. The *Brassica* blossoms must be cross-pollinated—fertilized by pollen from another plant. In fact, each plant's stigma not only can chemically recognize its own pollen, but has a way to physically prevent its own pollen from reaching its ovule. Cross-pollination ensures that the plant genes are well crossed.

The *Brassica's* yellow pollen grains are so heavy and sticky that they can not be picked up by the wind. Somehow this immobile plant must get some pollen from another plant in order to produce fertile seeds. It must also have a way to contribute pollen to other plants. Happily, a remarkably well adapted pollen vector has evolved: the worker bee.

From the bee's perspective, the flower represents food—nectar and pollen. Attracted to the flower by its bright yellow color, the worker bee lands on a petal and thrusts her head deeply into the flower to reach the sweet nectar, sucking it up through a straw-like tongue. As the bee does this, her body also brushes past the flower's anthers, and her hairs trap pollen. The bee also brushes against the stigma, which is sticky. The sticky stigma picks up other flowers' pollen trapped in the bee's hairs. Unwittingly, while foraging for food the bee has also cross-pollinated the plants. Figure 11-1 on pg. 70 shows a bee pollinating a *Brassica* plant.

Before the bee flies back to the hive, she uses brushes on her midleg to collect excess pollen from her head and thorax. Then she places the pollen in specialized pollen baskets, located on the hind leg, for transport back to the hive. There it is consumed by the bees for its protein, fats, vitamins, and minerals.

The nectar gathered from the plant is stored by the bee in her crop, a nondigesting honey stomach. When the crop is full, the bee regurgitates the nectar in it into a storage cell in the hive. The nectar is the bee's source of carbohydrates. Through evaporation and the action of enzymes, the nectar becomes honey.

Bee and flower: each has benefited from the relationship. Each has provided the other with vital necessities. Each depends on the other to survive.

Materials

For each student

   1   bee stick (made in previous lesson)
       Plants with open flowers

For every two students

   1   hand lens

For the class

   1   overhead transparency, "Bee Pollinating a *Brassica* Flower" (Figure 11-1)
   1   projector and screen

Preparation

1. Read the **Background** information above.

2. Set up the projector and screen.

Procedure

1. Pollination should be repeated each day between Day 12 and Day 18. After the first experience or after you are satisfied that the students are cross-pollinating correctly, they can continue the job at unscheduled times during the day.

2. Direct children to pick up their plants, bee sticks, and hand lenses at the distribution station.

3. Tell the students to use the bees to transfer pollen from the blossom on one plant to the blossom on another plant. They should pollinate every blossom that is open by rotating the bee gently. Circulate and remind them to cross-pollinate by taking pollen from one plant and transferring it to another.

4. Stop the pollination process periodically and ask the children to observe with their hand lenses. Look for:

    a. pollen on different parts of the plant

    b. pollen trapped in the bee's hairs

5. After the children have finished pollinating, have them clean up. Then project the transparency and allow the students to observe it for a short time, perhaps during the cleanup period.

6. Initiate a discussion on pollination. Help students come to the conclusion that both the bee and the blossom benefit.

    Ask: "In real life, what attracts the bee to the flower?" (Color and scent.) Then lead a discussion about how the bee reaches the sweet nectar in the bottom of the flower. Discuss how the bee squeezes between the anthers and the stigma to reach the nectar with her straw-like tongue. Find out if students noticed the yellow pollen grains caught in the hairs of the bee's body. From this discussion, help students understand that the bee gets food from the flower.

7. To help students see what the flower gets from the bee, look at the transparency showing both the male and female parts of the flower (see Figure 11-1). Point out the female parts: the pistil with the sticky stigma at the end. Then point out the male parts: the anthers on stalks called filaments. The anthers hold the pollen.

    Ask: "How do you think the male part (or the pollen) of one plant can get to the female part (or stigma) of another plant?" The answer is bees, of course!

Final Activities

Have students read *The Bee and the* Brassica: *Interdependence* on pg. 35 in the Student Activity Book (pg. 68 in the Teacher's Guide).

Extensions

1. Have students do extra reading on interdependence between bees and flowers. If possible, show a filmstrip or video on the subject. See **Appendix D** for suggested products.

2. In Lesson 4, the class removed extra plants from their planters and placed them into a class container. Now is the time to label them "Control Plants: Do Not Pollinate." Hold a brief discussion during which you ask students to predict what will happen to the unpollinated plants. Record the students' ideas. Save the predictions until Lesson 16, when the students harvest and thresh their crop.

Evaluation

It is important to assess comprehension of pollination and interdependence before students go on to model-building and dramatizing in the next lessons. Students should have learned the following:

- the basic vocabulary of plant reproduction, especially pollen, anther, and stigma

- the bee's role in cross-pollination

- an understanding of why the plant and the bee benefit from their relationship

**Reading
Selection**

1. Check the levels in your water tanks and refill if necessary.

2. After pollinating for the last time on Day 18, have the children pinch off any unopened buds.

The Bee and the *Brassica*: Interdependence

Bees and *Brassica* plants need each other in order to live. Each one takes something from the other and gives something in return. You might say that they have a real partnership.

Why does a flower need a bee? The main reason is so that the flower can make seeds. The *Brassica* flower holds both the male and the female parts of the plant. The male parts, the filament and anther, produce the pollen, which looks like fine yellow powder. Pollen must travel to the female parts, the pistil and stigma, of another flower on a different *Brassica* plant. Unless the pollen from one plant can reach another plant, no new seeds will form. Then, no new *Brassica* seedlings will grow.

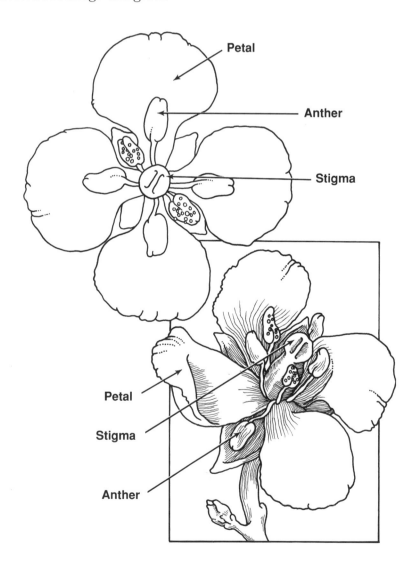

So it is very important that the pollen get from one plant to another. But the problem is that the pollen is sticky and cannot easily travel in the wind. How can the pollen travel? That's where the worker bee comes in. With its bright yellow color and sweet perfume, the flower lures the bee and offers not only one but two kinds of food: nectar and pollen.

The bee's body is covered with feathery hairs. As the bee dips her head into the flower to sip the sweet nectar deep inside the blossom, her hairy body rubs against the anthers holding the pollen. Her body traps some of it. When the bee flies off to the next flower, some of the pollen on her body sticks to the stigma there.

Now the bee has done her job. The bee has collected two kinds of food from the flower. At the same time, it has carried pollen from one flower to another. New seeds will form. Soon new flowers will bloom.

Figure 11-1

Bee pollinating a
Brassica *flower*

Observing Pods
(Day 17 to 35)

Overview

In nature, everything has its time and place. The flower petals attract bees, allowing pollination to take place. Now the flower will wither and die, and the seeds will develop. In this lesson, students will observe this miraculous process.

Objectives

■ Students observe the development of the fertilized pods between Day 17 and Day 35.

■ Students record their observations by drawing, writing, and graphing.

Background

Soon after pollination and fertilization, the flower will begin to change. The petals fade, wither, and fall. The ovules, or eggs, inside the ovary are developing into seeds. The pods, which are actually enlarged ovaries, grow and swell. The process continues until about Day 36, when the plant is removed from the watering system so that the pods can dry and ripen.

During this time, there will be very little upward growth because the plant is expending its energy on seed production. Continued measurements will confirm this for students.

Plan on having students observe and record pod growth on about Days 17, 24, and 31.

Materials

For each student
   1   plant with pods developing
   1   sheet of centimeter graph paper (see **Appendix F** for black line
          master)
   1   **Observation Sheet** (see the Observation Sheet from Lesson 6, pg. 43)
   1   toothpick (optional)
   1   student notebook

For every two students
   1   hand lens
   1   pair of forceps (optional)

For the class
Life Cycle Cards 8 and 9

Preparation

Duplicate additional copies of the graph paper from **Appendix F** and the **Observation Sheet** from Lesson 6, if necessary.

Procedure

1. Tell children to observe their plants with a hand lens and to focus on the flowers. Many will be dead and will have been replaced by pods. Allow sufficient time for questions and comments.

2. Open a discussion about the changes students observed. Ask students what is happening to the petals. Students will probably say that the petals are fading from yellow to white, wilting, shriveling and drying up, and falling off. The reason is that the flowers have been pollinated and do not need their bright colored petals to attract bees anymore.

    Then discuss what has happened to the anthers. Observe that some may still be attached to the plant, while others may have dried and fallen off.

    Finally, discuss how the pistil looks now. Students may describe it as larger or longer, swollen, lumpy, or bumpy. Explain that the new development is called the seed pod. It will continue to grow until you take the plants away from the water on Day 36. (Keep preparing them for this!) Ask them what they think is inside the pod. Talk about the answer: seeds.

3. Tell students that they will continue to observe their plant and to record its progress in their notebooks. Distribute the observation sheet and graph paper. Ask them to draw the plant and describe it in words on the observation sheet. Have them measure the plant's height and record it on their growth graph. Students will notice that the plant is no longer growing upward.

4. Remind them to make observations at least once a week until Day 35.

Final Activities

1. Display Life Cycle Cards 8 and 9 showing pod development (see Figure 12-1). Ask students to focus on the variety of developmental stages present on the same plant at the same time. The buds, the youngest part of the plant, will be closest to the top, while the most developed pods will be found lower down on the stem.

2. Remind students to record their observations of this stage of development.

Check water levels and refill tanks up until Day 35.

On Day 36, empty water tanks and allow the plants to dry out. Seeds will ripen and be ready for harvest by Day 42.

Extensions

1. Allow interested students to dissect a seed pod. Use forceps and toothpicks to open up the pod and look inside.

2. Feature a pod collection in the learning center. Good pods to examine are fresh beans and peas, but even canned string beans or wax beans are worth looking into.

Figure 12-1

Brassica *Pods*

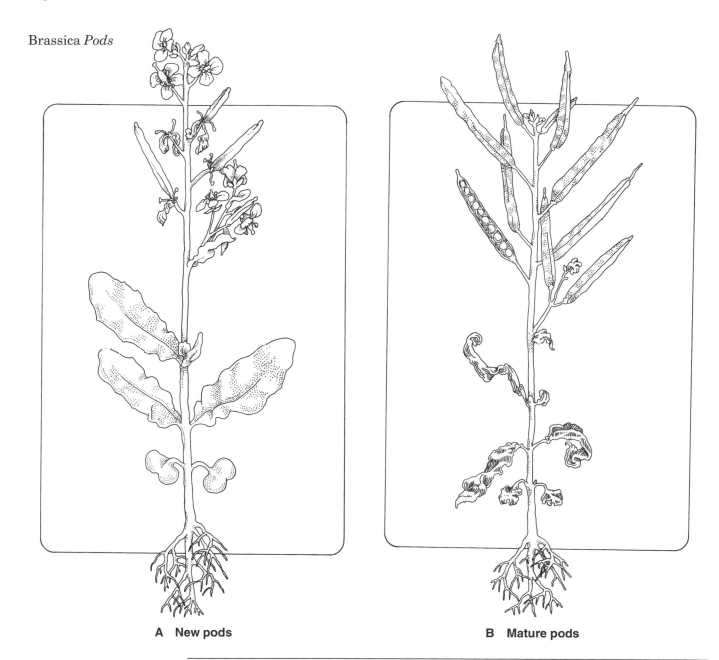

A   New pods   B   Mature pods

3. Challenge students to keep measurements of the length of a developing pod. This is a more difficult task, requiring careful handling of the plant and the ability to measure precisely.

Evaluation

1. Continue to monitor students' progress in making observations that are clear, complete and accurate. Over time, their drawings of pods should show increased size and more pronounced bumps.

2. Assess student progress in graphing independently.

Making a *Brassica* Model

Overview

At this point in the unit, students have studied the *Brassica* flower and the bee in depth. During the next two lessons, students will have a chance to apply what they have learned by making models of both the flower and the bee. Model-building offers an opportunity to incorporate a number of other skills, such as planning, measuring, cutting, assembling, and organizing parts. The more proficient students are at these skills, the more accurate and aesthetically pleasing their models will be.

Objectives

■ Students apply skills they have learned to construct an accurate model of the *Brassica*.

■ Students work together on a group project

Background

Making models is a combination of science, technology, mathematics, and art. A model is a useful vehicle for illustrating small structures, such as those found in a flower or on a bee. A set of flower and bee models also can be used to dramatically demonstrate their mutually beneficial relationship.

By now, students have had a great deal of practice making observations. Those skills will come in handy as they try to replicate reality in the form of anatomically correct models. As much as possible, encourage students to design and build their own models based on their own observations. The plans described in this lesson are meant to serve only as starting points, not as blueprints.

There are two different options suggested for making *Brassica* models in this lesson. Option A is the easier of the two. Option B is much more challenging. Also provided are two separate lists of materials and directions, a style that is different from the usual format of the lessons in the unit. All the other elements in the lesson are the same.

The models constructed in this lesson can be used in a dramatization of cross-pollination and interdependence. (See Lesson 14 for a sample script.) For this reason, the size of the models is important. You want them large enough to be seen across the classroom, yet small enough for students to manipulate. Sets of models (ideally, one bee to two flowers in order to show cross-pollination) should correspond roughly in size.

Procedure

1. Group the students in cooperative work teams. Give them time to brainstorm about:

   ■ what the model should look like

   ■ what materials they will try to collect (Encourage recycling!)

   ■ what each member of the team will be responsible for

2. Hold a group discussion to help students finalize plans for their models. Encourage them to develop their own models. To give everyone a head start, two sets of instructions for making *Brassica* flower models are given below. Option A includes instructions for making only the *Brassica* blossom. Option B includes instructions for making the whole plant.

OPTION A

Materials

For petals

   4 yellow styrofoam meat trays
      OR
   1 plastic milk jug painted yellow
      OR
      Cardboard painted yellow

For anthers and pistil

   1 styrofoam meat tray
      OR
   1 plastic milk jug
      OR
      Cardboard

To hold pistil and anthers

   1 styrofoam or paper cup
   4 brass fasteners
   1 stapler
   1 pair of scissors
      Tempera paint and brush

DIRECTIONS

1. Trace and cut out four yellow petals. See Figure 13-1.

2. Attach the petals to each other. Staples work well for thin materials, but brass fasteners are better for cardboard or styrofoam. Figure 13-2 shows how to fasten the petals together.

3. Trace and cut one pistil, four tall anthers, and two short anthers (see Figure 13-3). One idea is to glue sand to the anthers to simulate pollen, then to use glitter to make the stigma glisten.

4. Staple the anthers to the pistil.

Figure 13-1

*Petals to
trace and cut*

Figure 13-2

Attaching the petals

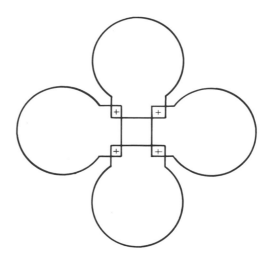

5. Cut a slit in the bottom of the styrofoam or paper cup. Poke the bottoms of the pistil and anthers through the slit. Figure 13-4 shows you how.

6. Force the cup into the space in the center of the four petals. The blossom is complete. Children may hold it by the cup to manipulate it during the dramatization. This type of model needs support, such as an empty coffee can, in order to stand.

OPTION B

Materials

For petals
   4 yellow styrofoam meat trays
      OR
      1 plastic milk jug painted yellow
      OR
      Cardboard painted yellow

For anthers
   6 styrofoam packing "peanuts" painted yellow
      OR
   6 yellow jelly beans

For filaments to support the anthers
   6 pipe cleaners
      OR
   6 soda straws
      OR
   6 popsicle sticks
      OR
   6 coffee stirrers

Figure 13-3

Trace and cut one pistil, four tall anthers, and two short anthers

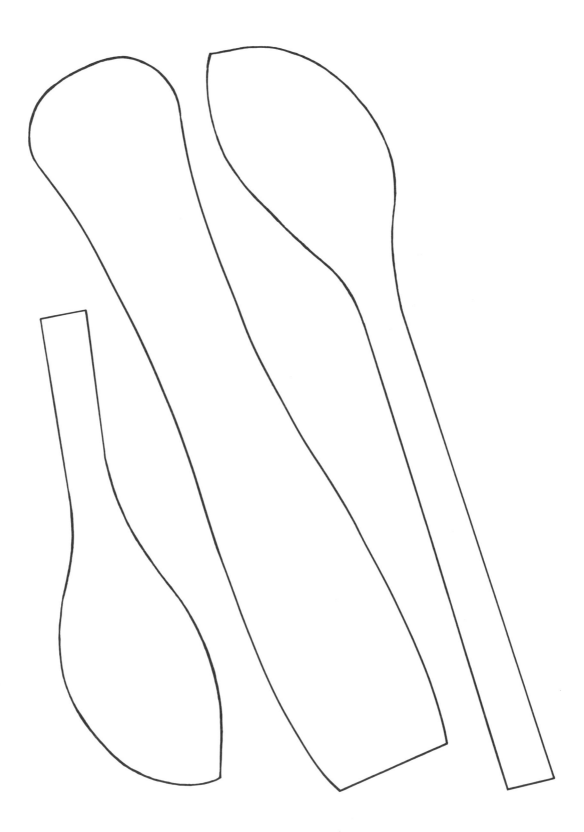

Figure 13-4

*Putting the
flower together*

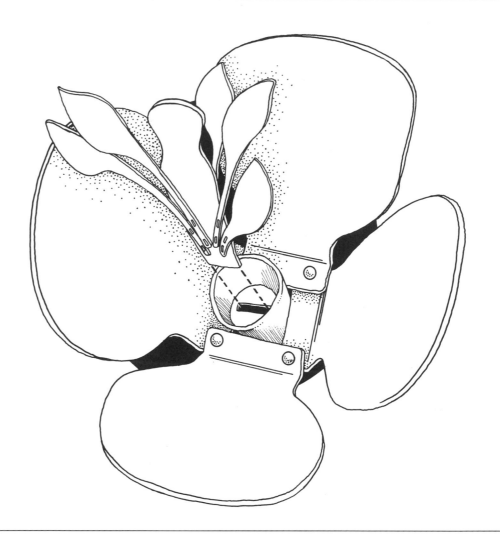

For leaves and sepals

    2   green plastic 2-liter soda bottles
       OR
       Cardboard painted green
       OR
       A milk jug painted green

For the pistil

    1   cardboard tube

For the stem

    1   dowel or sturdy twig

For fastening parts to the stem

       Modeling clay
       OR
       Styrofoam balls with a center hole

DIRECTIONS

1. Draw and cut out the following parts and paint them the appropriate colors:

    4 petals

    4 sepals

    3 to 5 true leaves

    2 seed leaves

2. Paint the six anthers yellow. Poke four long and two short filaments into the anthers to support them. Use pipe cleaners, soda straws, popsicle sticks, or coffee stirrers for the filaments.

3. Fasten all the parts to the stem. Use lumps of modeling clay or styrofoam balls with a center hole cut out to fit the dowel. Use this as the base for attaching parts to the stem.

4. Now the model is complete (see Figure 13-5). It can be either hand held or stood up in a coffee can of sand for support.

Figure 13-5

Fully assembled model plant

Final Activities

1. Provide space for the children to exhibit their models.

2. Start organizing the class into groups for building model bees, which will take place during the next class. Encourage students to review their bee drawings and to take time to observe the real bees in preparation for model-building.

Making a *Brassica* Model / **81**

Extensions

Have students critique each others' models. Provide some guidelines, such as:

■ Does the model show the correct numbers of petals and anthers? Is only one pistil shown?

■ Are the parts in the right places?

■ Has the model been done neatly? Is it attractive? How much effort does it show?

Begin planning ahead for a class performance of a dramatization entitled "The Bee and the *Brassica*" (see Lesson 14, **Extensions**, for more details). With the class, decide when it will take place and who to invite to the show.

Evaluation

Evaluate the models using the same criteria as suggested for the student critiques under **Extensions**.

Making a Bee Model

Overview

During this lesson, students will make an anatomically correct model of a bee. This activity draws on what students have learned about bees through observing and using them as tools for pollination.

Objectives

■ Students construct an accurate model of a bee.

■ Students work together on a group project.

Background

Ideally, students will create their own bee models. Children are far more interested in and invested in models springing from their own imaginations. Realistically, however, it is not always possible to expend the time and effort for original creations. So suggestions for two kinds of bee models are described below.

The first is a stick puppet, and is the easier of the two (see Figure 14-1). The second is a three-dimensional, self-standing model requiring a great deal more skill. There also are two separate lists of materials and directions, a style that is different from the usual format. All the other elements of the lesson are the same.

Procedure

1. Group the students in cooperative work teams, and give them time to brainstorm about:

   ■ what the bee model should look like

   ■ what materials they should try to collect (continue to suggest recycling)

   ■ what each member of the team will be responsible for

2. Have a class discussion about the bee models. Talk about how the bee should relate to the flower models in size. Help students learn from the experience they gained in building and evaluating the flower models.

3. Encourage children to create their own designs. If this is not possible, you may want to use one of the plans below. A complete stick puppet is shown in Figure 14-1.

Figure 14-1

Stick puppet

OPTION A

Materials   *For each team*

For the body
1   copy of the bee drawing (Figure 14-2) or a freehand drawing

For backing
Stiff paper
OR
Poster board
OR
Cardboard

For the handle
Paint stirrer
OR
Ruler
OR
Heavy cardboard

For wings
Clear plastic 2-liter soda bottle
OR
Transparency film

For the stinger
Cardboard
OR
Popsicle stick

Figure 14-2

Bee body to trace and cut

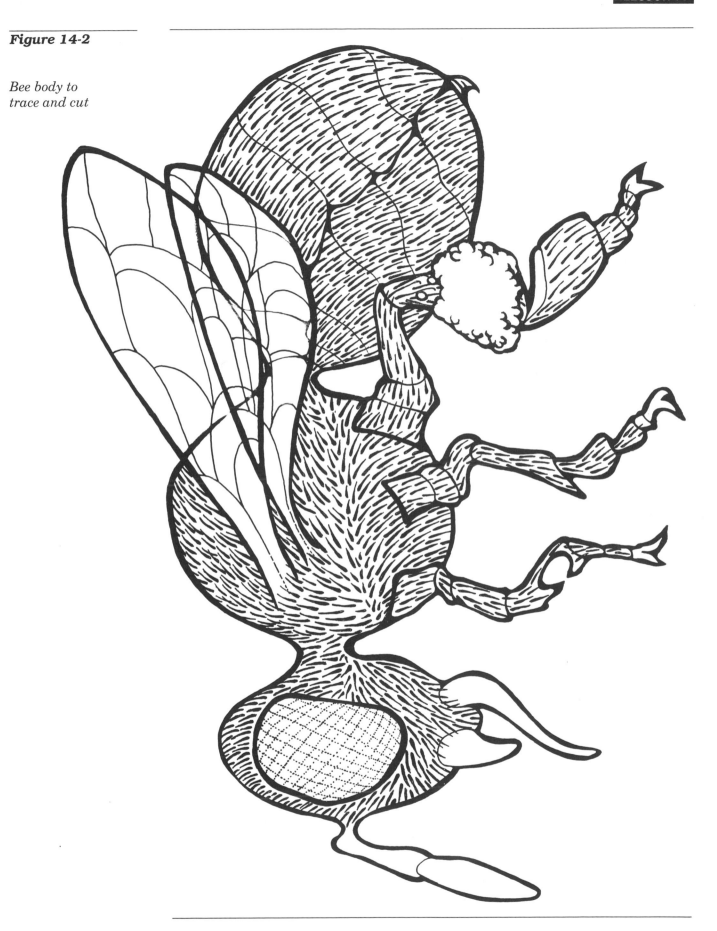

For hair
>   Fuzzy fabric scraps
>   OR
>   Yarn

For the tongue
>   Pipe cleaner
>   OR
>   Soda straw

For pollen basket
>   Sponge pieces
>   OR
>   Cotton balls
>   OR
>   Fabric scraps

Other supplies
>   Glue
>   Colored markers
>   Stapler
>   Scissors

DIRECTIONS

1. Pass out an outline drawing of the bee. Tell students to use it as a base upon which to build.

2. Glue the bee drawing to a heavier piece of paper used as backing.

3. Cut out the bee glued to the backing.

4. Color the bee with crayons or markers.

5. Then get creative. Add all sorts of interesting embellishments. For example, cut four transparent wings from plastic soda bottles and staple them to the bee. Add on a stinger or "hair." Staple on a soda straw or a pipe cleaner for a tongue. Glue sponge pieces or cotton balls to the rear legs to represent pollen baskets.

6. Attach the bee to the stick. A stapler will probably work fine for most materials.

OPTION B

Materials

For each team

For the body
>   1   styrofoam block, 2" x 2" x 8"

For the wings
> Clear plastic 2-liter soda bottle
> OR
> Transparency film

For the stinger
> toothpick

For legs
> 6 black pipe cleaners

For pollen baskets
> Cotton balls
> OR
> Sponge pieces

For the tongue
> Straw
> OR
> Pipe cleaner

Other supplies
> Serrated plastic knife
> Sandpaper or rasp
> Glue
> Black marker
> Tempera paint and brush
> Toothpicks

DIRECTIONS

1. Mark off the styrofoam block into three sections of 1½ inches, 2½ inches, and 4 inches in length. Mark these segments all the way around the four sides of the foam block. Figure 14-3 shows how to do this.

Figure 14-3

Marking the styrofoam block

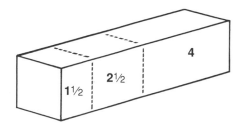

2. With a serrated knife, cut out V-shaped wedges at the marks on each edge. Figure 14-4, A, shows how to do this.

3. Then cut out V-shaped wedges to connect the cuts you have already made. This will divide the styrofoam block into a rough head, thorax, and abdomen. Figure 14-4, B, shows this.

Figure 14-4

Shaping the body sections

A   B

4. Use sandpaper or a file to refine the shape of each section on the body.

5. With a marker, draw in the three simple and two compound eyes.

6. Paint the body with tempera paints.

7. Cut the pipe cleaners into:

   - five 3-inch pieces for front legs, antennae, and tongue

   - two 4-inch pieces for middle legs

   - two 5-inch pieces for hind legs

8. With a toothpick, punch holes at an angle to insert the pipe cleaner legs, antennae, and tongue. Figure 14-5 shows what this looks like.

Figure 14-5

Side view of the model bee's body

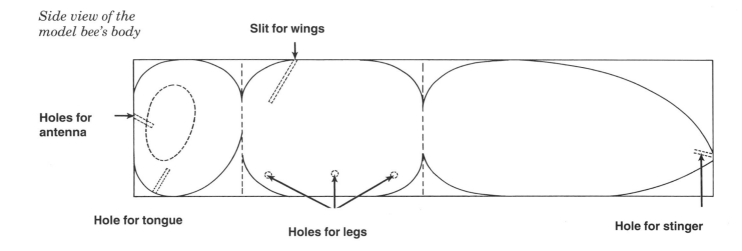

Slit for wings

Holes for antenna

Hole for tongue

Holes for legs

Hole for stinger

9. Cover the pipe cleaner tips with glue. Insert the glued tips into the holes. Allow the glue to set while you work on the wings.

10. Cut two of each kind of wing (total four wings) from the clear plastic. Draw in the wing veins with a black marker. Assemble the wings into two pairs, with the smaller wing beneath the larger and the tabs together. Figure 14-6 shows the wings.

11. With a knife, make slits in the thorax for the wings. Be sure to make the cuts wide enough and deep enough for the wing tabs. Then cover the wing tabs with glue and insert one pair of wings into each slit.

Figure 14-6

Wings

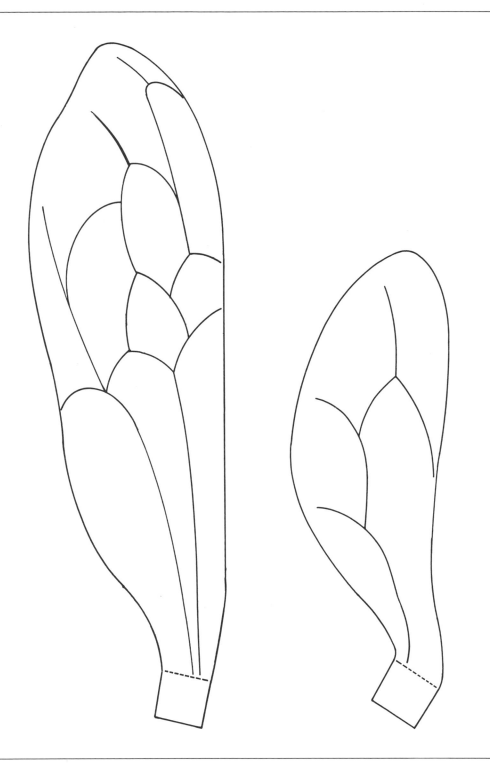

12. Punch a hole through the cotton balls or pieces of sponge and insert them onto the hind legs. These are the pollen baskets.

13. Bend the legs and antennae as shown in the illustration of the fully assembled bee model (Figure 14-7). The completed model will stand by itself.

Figure 14-7

Completed bee model

Pollen basket

Wings

Three simple eyes

Two compound eyes

Antenna

Legs

Tongue

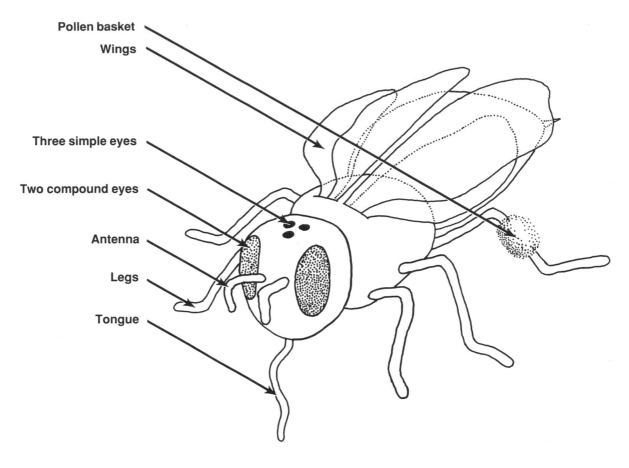

Final Activities

1. Provide space for the children to exhibit their models.

2. Plan to use the models to give a performance of "The Bee and the *Brassica*," a true-life drama of interdependence. Suggestions about how to do this are explained under **Extensions**, below.

Extensions

Have students act out the natural drama that takes place between the bee and the *Brassica*. A spontaneous "round-robin" narrative is not only great fun, but also allows the teacher to assess how much the students have learned about cross-pollination and interdependence. Look for the use of appropriate vocabulary and the correct sequencing of events as the children tell the story in their own words. Below is a description of how the round-robin works.

Select a group of four to six students to hold the flower models and place them "on stage" in front of the room. Select one or two students to manipulate the bee models. Ask them to wait off to the side until it is time for them to come in. Tell these two groups that they will act out the story as other members of the class tell it.

Tell the rest of the class that they will all be involved in a round-robin narration of the story of interdependence and cross-pollination while the models act it out.

Here is how it works. Ask one child to begin the story. At a crucial point say "Buzz" to another child, who will continue the story where the first child left off.

Before beginning the activity, run through an example with the class. Make sure students understand the method before proceeding.

Below is an example of a round-robin script to help get the idea.

Student 1: "One warm and sunny summer day, a bee was flying over a meadow when suddenly she saw some bright yellow flowers swaying in the breeze."

(At the same time, the students working the bee models are circling the area and the flower models are swaying enticingly.)

Buzz to Student 2.

Student 2: "The bee landed lightly on the yellow petal..."

(Bees land on petals.)

Buzz to Student 3.

Student 3: "...and forced her way deep into the flower to reach the sweet nectar."

(Bees enter flowers.)

Buzz to Student 4.

Student 4: "Meanwhile, back at the thorax, the bee brushes against the anther of the flower and some pollen gets caught on her hairy body."

Buzz to Student 5.

Student 5: "Now the bee has collected both nectar and pollen from this flower and is ready to take off. While backing out..."

Buzz to Student 6.

Student 6: "...her body brushes against the sticky stigma, and some pollen sticks to the stigma."

(Body and stigma brush.)

Buzz to Student 7.

Student 7: "But the bee's honey stomach is not full of nectar yet, so she flies off to another flower."

(Bee flies off to another flower.)

Now the story gets repeated with the second flower. Only this time it is important to emphasize that cross-pollination is taking place. Keep the narration going until all the children have participated in some way, or until your patience wears thin. Then thank all the participants, and give yourselves a round of applause.

As a further extension, have students write the script for a dramatization of pollination and interdependence. These skits could be presented to an audience of parents or classmates.

Evaluation

1.   The bee models can be evaluated using the following criteria:

   ■   Does the model show the correct numbers of parts?

   ■   Are the body parts in the right places?

   ■   Can the student explain the model to you?

   ■   Has the model been neatly done? Is it attractive? How much effort does it show?

2.   As you watch their performances or read their scripts of the drama, evaluate students on:

   ■   Correct use of appropriate vocabulary.

   ■   Accurate sequencing of events.

   ■   An understanding of the process of cross-pollination and the interdependent relationship of the bee and flower.

LESSON 15	# Interpreting Graphs

Overview

Students have had a great deal of practice constructing graphs during this unit. During this lesson, they are asked to interpret information on two different graphs. This is a good review of graphing for students, as well as a diagnostic tool.

Objectives

- Students interpret information on two different graphs.
- Students apply math skills to reading graphs.

Background

This lesson provides you with a good opportunity to see how well students can read graphs.

Materials

For each student

   1 student notebook

Procedure

1. Begin the lesson by commenting that you have noticed progress in the students' ability to make graphs of their plants' growth. Tell them that today they will practice getting information from graphs.

2. Tell students to open their Student Activity Books to pg. 44 in Lesson 15 and to spend a few minutes studying the graph (Figure 15-1 in both the Student Activity Book and the Teacher's Guide). Tell the students that the graph is a comparison of how long it takes for five different plants to make seeds. Then ask questions about the different elements of the graph, such as what the two axes represent and what plants are being compared. Notice, too, that the graph is written in increments of 5 days.

3. Next, ask students to use the graph to answer the following questions (also found in **Find Out for Yourself** on pg. 43 in the Student Activity Book). Tell them to record the answers in their notebooks.

   a. What is the title of the graph?

   b. How many days does it take for Wisconsin Fast Plants™ to develop seeds?

Figure 15-1

Graph of plant life cycles

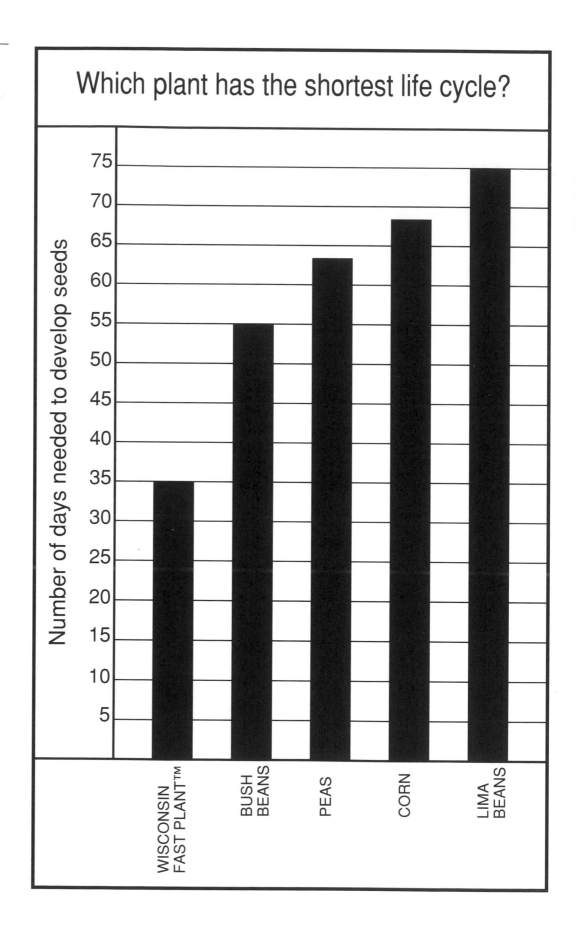

Which plant has the shortest life cycle?

c. After Wisconsin Fast Plants™, what is the next fastest plant to develop seeds?

d. How many days does it take for pea seeds to develop? Notice that the bar stops between the 60-day and the 65-day marks. Make a good estimate of which day this means.

e. Lima beans need 75 days to develop seeds. Corn only needs about 68 days. How many more days does it take for lima beans to develop seeds than corn?

4. When everyone has completed work on the plant graph, review the answers together. Make an effort to pinpoint and correct problems before telling everyone to turn to Helen's Graph on pg. 46 in the Student Activity Book (see Figure 15-2).

5. Give students time to read the paragraph that precedes Helen's Graph.

> Every fourth year, it was Helen's turn to spend the summer at Grandma and Grandpa's farm. Even before she had time to unpack, Grandma would always say "Let's see how much you've grown since the last time you were here." Then the three of them would stroll out to the barn together. Right next to Chester the horse's stall was a smooth plank wall. This was Helen's measuring place. Helen would stand up straight with her heels to the wall, and Grandpa would draw a line across the plank to mark her height. Each time they measured, Helen would move over one plank to the right. And she always wrote her age under the measurement. By the time she was 20 years old, that wall looked just like a graph!

6. Say to the class, "Use Helen's Graph to answer the questions in the Student Activity Book. Record the answers in your notebook. When everyone has finished, we will discuss the answers together."

a. Helen was 45 cm long at birth. By the time she was ready to spend her first summer in the country at age 4, she was more than twice that tall. How tall was Helen at age 4?

b. Grandpa was big on safety. He insisted that no one could ride Chester alone unless that person was at least 125 cm tall. When she was only 4 it seemed like a good rule. But by the time she was 8, Helen was dreaming of riding solo. Did she get her wish that summer when she was 8?

c. Grandpa had another rule. You had to be at least 150 cm tall to drive the tractor. How old was Helen when she was first allowed to drive the tractor?

d. At 12, Helen was right in the middle of her adolescent growth spurt. How much taller is she at 12 than she was at 8?

e. Grandma and Helen stood back to back at the end of her sixteenth summer and discovered that they were exactly the same height, 165 cm. Do you think they were still the same height when Helen was 17 years old? Give a good reason for your answer.

Figure 15-2

Helen's graph

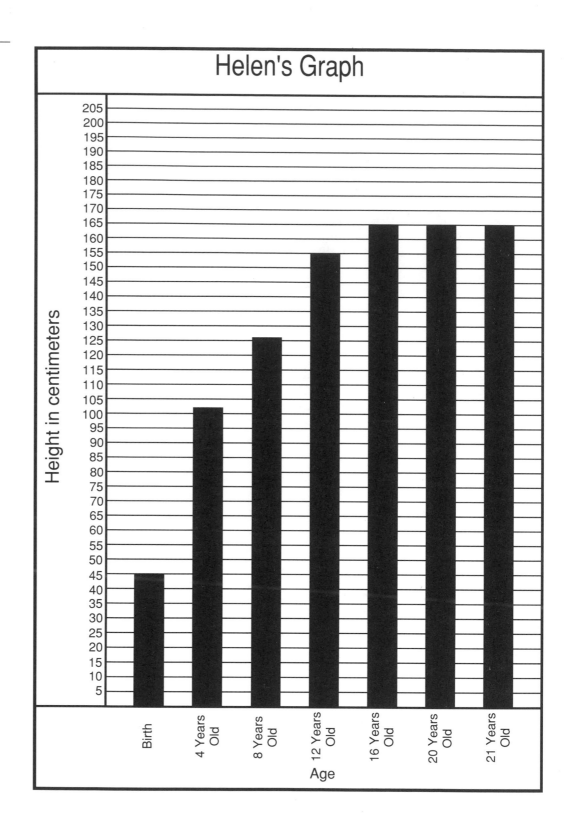

Helen's Graph

Final Activities   Discuss the questions about Helen's Graph. Help students think through
their answers and give reasons for them. Offer extra help to students having
trouble. For further practice, assign some of the activities in the section below.

Extensions

1. Many families keep growth graphs. Ask students to bring in data about
    their own growth and to record it on graph paper.

2. As a class, collect data about yourselves and graph it. Subjects might
    include:

    ■ Birthdays. How many people in the class have birthdays in different
       months of the year?

    ■ Favorite school lunch. Pick three favorites and poll the class.

    ■ Eye color

- Number of siblings
- Favorite TV show
- Bed time
- Pets

Evaluation

Check for progress in these areas:

- The ability to read quantities represented on the graph.
- The ability to estimate quantities that fall between the whole increments represented on the graph.
- The ability to make computations based on information gleaned from the graph.
- The ability to identify the elements of a graph and understand what they mean.

| LESSON 16 | # Harvesting and Threshing the Seeds |

Overview

At the end of the unit, it is satisfying to harvest, or cut, the crop and separate out, or thresh, the seeds. This completes the life cycle of a plant—from seed to seed. Students will be eager to discuss what they have learned and to discover how their new knowledge can lead to new questions and experiments.

Objectives

- Students harvest and thresh the seeds.

- Students count the seeds and compare that number with the original number of seeds planted (8) to determine their profit or loss.

- Students think about additional questions they have about plants and experiments that might help answer them.

Materials

For each student
- 1   tray
- 1   envelope for seed storage
- 1   pair of scissors (optional)
- 1   paper cup (optional)
- 1   quad of dried plants
- 1   student notebook

Preparation

Distribute materials.

Procedure

1. To begin, have children retrieve their plants and spend a few minutes observing their dried-out condition. Ask: "What do you observe that is different about your plants today?" Have students make observations using their senses. They might say that the color was green and is now brown. Or that the texture was once soft and supple and is now brittle and hard. Finally, they might notice that the pods now make a "snapping" sound. Have them record these observations in their notebooks.

    Expect to hear comments like "My plant is sad" or "I hate it that my plant died." It is important to recognize that the children have become very involved with their plants. It is equally important to point out that, although you recognize these as legitimate feelings, they are feelings and not observations.

2. Instruct students to place their planter on the tray and to harvest the pods either by snapping them off with their fingers or by cutting them with scissors. Ask: "How many pods have you harvested from each plant? What is the total number?" Instruct students to record this in their notebooks.

3. Tell children to gently roll each pod between their hands and over the tray (and into a cup if available) to thresh, or separate, the seeds from the pods.

4. Have students count all the seeds and record the number. Place seeds in an envelope for storage. Have students label the envelope with their name and date of harvest. See suggestions of what to do with the seeds in the section entitled **Extensions**. If you plan to keep the seeds over a period of months, store them in the refrigerator in an airtight container such as a jar or a plastic bag.

5. Compare the yield from this harvest to the original number of seeds planted. Do the students have more or fewer seeds than they did when they started? Why? Have students record their findings.

Final Activities

1. Display the nonpollinated plants that were set aside after Lesson 4. The differences between the two sets of plants should be quite obvious. Lead students to draw a conclusion about the importance of pollinating.

2. What other questions do the students have about plants? Encourage them to share these questions. New questions could lead to new experiments with plants, further reading in trade books, and group or individual projects using the seed supply.

Extensions

1. Discuss the world of agriculture. How does a farmer make a "profit"? Challenge the children to find out what a "combine" combines.

Figure 16-1

Combine

What would happen if farmers could raise "fast corn"? What if a "fast weed" developed?

2. Give interested students a chance to replant and conduct individual experiments. Since Wisconsin Fast Plants™ do not need a dormant period, the seeds may be replanted immediately or stored in a cool dry place for use later. Brainstorm about what conditions you could change to see how that condition influences growth and yield. For example:

- What would happen if the plants only got 12 hours of light per day instead of 24?

- How tall would the plants grow on twice the amount of fertilizer? How tall would they grow on half the amount?

- What if you grow two plants per cell instead of one? Why was it important to thin to one plant per section?

- Is a bee stick the best pollinator? What if we substituted a soft paint brush or a cotton swab?

Evaluation

1. Assess student progress in:

- using observable properties for their descriptions

- organizing data

2. This is also a good time to notice progress in students' ability to ask questions and suggest experiments to find the answers.

Post-Unit Assessments

Overview

- **Assessment 1** is a sequencing activity, where students put life cycle cards in the correct order.

- **Assessment 2** gives students a chance to draw a second drawing of a bee after observing and performing experiments.

- **Assessment 3** is a follow-up to the student brainstorming session about plants held in Lesson 10.

Objectives

- Students evaluate their own progress.

- Teachers evaluate student progress.

ASSESSMENT 1

Putting the Plant's Life Cycle in Sequential Order

This activity gives students a chance to show how much they have learned about life cycles by putting the Life Cycle Cards in the correct order.

Materials

For each student

    1   sheet of pictures of the *Brassica* life cycle (see pg. 105)
    1   pair of scissors
    1   **Activity Sheet 5** (optional)

Procedure

1. Distribute copies of the sheet that shows stages of the life cycle of the *Brassica*, copies of **Activity Sheet 5**, if you are going to use it, and the scissors.

2. Instruct children to cut carefully on the lines to separate the pictures.

3. Ask the children to arrange the cards in the correct order in a straight line in front of them, beginning at the left. Encourage close observation of details.

4. When the children are satisfied that their sequence is correct, ask them to number the cards from one to nine and to put their name or initials on the back of each. This not only forces the children to review their sequence, but also helps prevent mix-ups. If you prefer, they also could paste the sequence down on **Activity Sheet 5**.

5. For the correct life cycle sequence, see pg. 107.

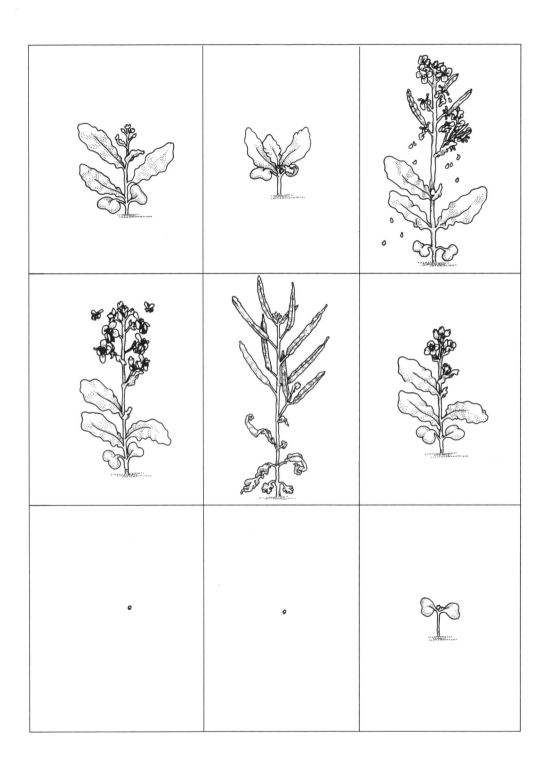

Name: _____

Date: _____

Cut out the cards. Put them in the right order. Then put the correct number in each box.

Name: _____

Date: _____

Cut out the cards. Put them in the right order. Then put the correct number in each box.

1

2

3

4

5

6

7

8

9

ASSESSMENT 2   Comparing Drawings

In Lesson 8, students drew a picture of what they think a bee looks like. In this assessment, students draw a second picture of a bee based on their observations.

Materials

For each student

    1   sheet of drawing paper
        Pencil
        Crayons or markers (optional)
        The child's own drawing of the bee done in Lesson 8

Preparation

1. Prepare a display area (bulletin board or clothesline). Title it something like "Look How Much We Have Learned About Bees."

2. Get out the original drawings done in Lesson 8.

Procedure

1. Tell the class that today they will see for themselves how much they have all learned about bees. Remind them of their previous drawings of the bee, which you have kept safe (and hidden). Tell them they will each compare their old drawing with today's work. Assure them that they will be pleasantly surprised.

2. Instruct the students to draw a bee with as many details as they can remember. Ask them to pay attention to the relative size of things, to the numbers of parts (such as legs, eyes, and wings), and to the location of parts.

3. When everyone is finished with the new drawing, pass out the old ones and let the children compare the two to evaluate their own progress.

ASSESSMENT 3   Follow-up on Student Brainstorming about Plants

During the early brainstorming session, students developed two lists: "What We Know About Plants" and "What We Would Like to Know About Plants." In this assessment students consider much they have learned.

Materials

The two student lists saved from Lesson 1.

Procedure

1. Display the two lists and analyze them with the students.

2. Look at List 1 "What We Know About Plants."

    a. Ask students to identify statements that they now know to be true without a doubt. Were there any experiences during the unit that confirmed these statements?

    b. Ask students to correct or improve those statements that need it. Have students give reasons for the corrections.

    c. Ask students to evaluate their progress. Ask them what other questions they would like to add to the list.

3. Look at List 2 "What We Would Like to Know About Plants."

    a. Tell students to go through the list of questions and pick out the ones they can answer now.

    b. Ask students to suggest ways to find out the answers to questions that were not answered.

4. Applaud students for their progress.

5. Encourage the class to go on looking for the answers to questions that were not answered in the unit.

APPENDIX B

Teacher's Record Chart
of Student Progress

Teacher's Record Chart of Student Progress for *Plant Growth and Development*

		Student
Products	Lesson 1: Activity Sheet 1, dry bean section completed	
	Lesson 2: Activity Sheet 1, soaked bean section completed	
	Lesson 3: Activity Sheet 2, record of planting	
	Lesson 4: Written observations and drawings of uprooted seedling showing roots, stem, and seed leaves	
	Lesson 5: Measurements of plant height in centimeters	
	Lesson 6: Written observations and drawings of seed leaves, true leaves, and buds	
	Lesson 7: Activity Sheet 3, measurements of plant height in centimeters and written observations of the growth spurt	
	Lesson 9: Activity Sheet 4, record of making a beestick and bee observations	
	Lesson 10: Written observations and drawing of the *Brassica* flower	
	Lesson 12: Written observations and drawings of seed pods	
	Lesson 12: Graph of plant height in centimeters	
	Lesson 15: Answers to questions on interpreting graphs	
	Lesson 16: Written observations of dried out plants	
	Lesson 16: Data on numbers of pods and seeds produced by own plants	
Specific Skills	Can distinguish predictions from wild guessing and can make reasonable predictions about plant growth and development	
	Can plant and take care of plants	
	Can observe changes in plants	
	Can keep written record of observations	
	Can record plant growth on a graph	
	Understands the reasons for thinning and transplanting and can carry out these two tasks	
	Understands what a growth spurt is and has watched plants going through this phase of development	
	Understands the basics of bee anatomy. Can identify the head, thorax, abdomen, hairs, pollen baskets, wings, eyes, antennae	
	Understands the basics of flower anatomy. Can identify petals, stigma, pistil, anthers, and pollen	
	Understands what is meant by pollination, and appreciates its importance to the production of seeds	
	Understands that there is an interdependent relationship between the bee and the flowering plant	
	Can create anatomically correct models of a bee and a *Brassica* flower, and can use them to explain pollination and interdependence	
	Can interpret information on a graph	
	Understands the concept of a plant's life cycle and can sequence it correctly	
General Skills	Follows directions	
	Records observations with drawings or words	
	Works cooperatively	
	Contributes to discussions	

APPENDIX C	# Introduction to Graphing

Overview

This optional lesson is provided for the class that needs it, either because the students have not yet studied graphing or because they need a review.

Objectives

- Students get an introduction to graphing.
- Students identify parts of a graph and interpret data recorded on a graph.

Background

This lesson should be used as an introduction to graphing and not as an introduction to the unit on Wisconsin Fast Plants™. It should be scheduled before the planting date because once the seeds are planted there is simply not enough time to teach the skill before students are expected to use it.

The unit assumes that the children know how to graph and can do it independently. The students get lots of practice applying graphing skills to a real-life situation.

If your class has already studied graphing, feel free to omit this lesson or to use it as a quick review before beginning the unit.

Materials

For the class

      Overhead projector and screen

1 set of transparencies, *Graphing*

Preparation

(~15 minutes)

1. Set up audiovisual equipment.
2. Preview the lesson and the transparencies. (Transparency masters are provided at the end of this appendix.)

Procedure

1. This lesson involves using five transparencies to help children see why symbols are useful and how graphs can help them organize information. Begin by projecting Frame 1. Ask students to count the objects on the screen. Because the objects are disorganized, children will have trouble counting them. Try to help them understand why it is difficult to count the objects.

2.   Remove Frame 1 and project Frame 2.

Ask: "How many of each kind of plant do you see?" Mention that they seemed to be able to count faster this time. Try to get them to see that the reason for this is that the plants are grouped (sorted, classified, organized) by kind.

Remove Frame 2. Ask: "Suppose we want to know how many more peanuts there are than lima beans. Can you think of a good way to show that in a picture?" Accept all students' ideas.

3.   Project Frame 3.

Have students compare the number of peanuts with that of lima beans. Help students recognize that the reason they were able to see this so quickly is that each kind of plant is arranged in a straight line. Tell students that this drawing is called a picture graph. But point out that drawing these pictures takes up a lot of room. Help them realize that using symbols to stand for plants is a much more efficient way to draw a graph. Discuss different symbols that could be used to stand for plants.

4.   Remove Frame 3, and project Frame 4. Explain that this graph uses seeds as symbols for plants. Point out that if there were a key, it would explain what the seed stands for.

5.   Leave Frame 4 on the screen, and put Frame 5 on top of it. Discuss the parts of a graph and why it is so easy to read. Point out that the graph title, the labels on the two sides (axes), the grid arrangement, and a key all contribute to readability. One way to make the graph even easier to read is by changing it into a bar graph.

6.   Keep Frames 4 and 5 on the screen, and place Frame 6 on top of them.

Point out that the bar graph shows the same information. Next, rotate the transparencies so that the bars go up and down rather than right to left. Ask: "Is the information on the graph different now?" Point out that even though you are reading in a different direction, the graph still tells you the same things.

7.   Remove all transparencies and turn off the projector.

Final Activities

Tell students that they will be doing a lot of graphing of plant growth over the next few weeks and that it is important that they understand it now before the unit begins. Ask a few review questions to clear up any trouble spots.

Extensions

Work together as a class to create a graph. Decide on a topic, what kind of data will be collected, and how to label the parts of the graph. Then ask the children to fill in the information on a blank sheet of graph paper. Here are some topics you might use:

- Shirt color. Find out how many children are wearing blue shirts, white shirts, pink shirts, others.

- Shoe style. Survey the group and find out how many are wearing sneakers, sandals, boots, others.

- Hours of TV watching. How much time does each child spend watching each day?

- Height. Measure everyone!

- Favorite after-school activity.

Evaluation

This lesson can help you diagnose any graphing problems early on while there is still time to do some extra teaching. Look for difficulties such as confusion about what the graph means, inability to identify the parts of a graph, adding information to the graph in the wrong place, or indicating the wrong quantity.

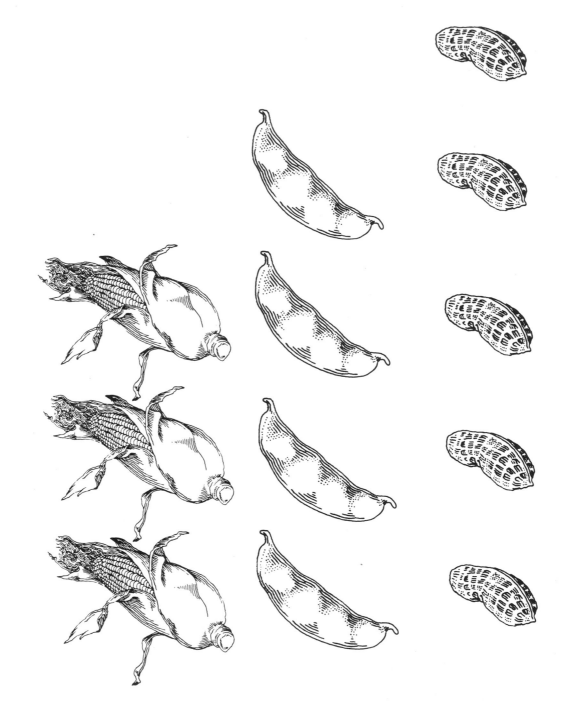

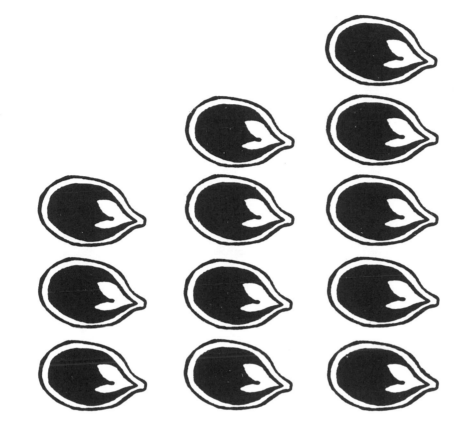

Seeds I Ate Today

Corn						
Lima Beans						
Peanuts						
	1	2	3	4	5	

Bibliography

Resources for Teachers

Dishon, Dee, and Wilson O'Leary. A *Guidebook for Cooperative Learning: Techniques for Creating More Effective Schools.* Holmes Beach, Florida: 1984.

A good practical guide to cooperative learning.

Johnson, David W., Johnson, Roger T., and Holubec, Edythe Johnson. *Circles of Learning.* Washington, D.C.: ASCD, 1984.

A good source of information about cooperative learning research.

Resources for Students

About Bees and Pollination

Carle, Eric, *The Honeybee and the Robber: A Moving Picture Book.* New York: Putnam Publishing Group, 1981.

An entertaining book with moveable parts.

Fischer-Nagel, Andreas, and Heiderose. *Life of the Honeybee.* Minneapolis: Carolrhoda Books, 1986.

Named an outstanding science trade book by the National Science Teachers Association, the book features excellent full-color pictures and a straightforward text.

Fleischman, Paul. *Joyful Noise*: *Poems for Two Voices.* New York: Harper & Row, 1988.

A slim volume of poetry to be read aloud by two people. The poem about honeybees is written from two points of view: the queen bee's and the worker bee's.

Lauber, Patricia. *From Flower to Flower.* New York: Crown, 1986.

Remarkable black-and-white photographs of different animals in the process of pollinating plants. Text is challenging.

Parker, Nancy Winslow, and Joan Richards Wright. *Bugs.* New York: Greenwillow Books, 1987.

> A lighthearted look at insects that manages to get in a lot of factual information.

About Plants and Gardening

Aliki. *Corn is Maize: The Gift of the Indians.* New York: Harper & Row, 1976.

> The story of how the Indians found and developed corn, and later shared this treasure with the new settlers.

Back, Christine, and Barrie Watts. *Bean and Plant.* Dobbs Ferry, New York: Sheridan, 1984.

> Full-color photographs show how a bean plant goes from seed to seed. Easy-to-read informative text.

Bellamy, David. *The Forest.* New York: Clarkson N. Potter, Inc., 1988.

Bellamy, David. *The Roadside.* New York: Clarkson N. Potter, Inc., 1988.

> Two beautifully and cleverly illustrated books showing plant and animal inhabitants of two different ecosystems. Both books touch on the theme of interdependence.

Brown, Marc. *Your First Garden Book.* Boston: Little, Brown & Company, 1981.

> More than twenty ideas for plant projects for indoors and outdoors, along with whimsical illustrations.

Cork, Barbara. *Mysteries and Marvels of Plant Life.* London: Usborne-Hayes Publishing Ltd., 1983.

> Every page is chock-full of plant oddities and unexplained marvels of nature. Full-color illustrations are both fun and informative.

Heller, Ruth. *The Reason for a Flower.* New York: Putnam Publishing Group, 1983.

> A sprinkling of text on brilliantly illustrated pages conveys a wealth of information about plants, pollination, and plant products. A real favorite.

Holley, Brian. *Plants and Flowers.* Burlington, England: Hayes Publishing Ltd., 1986.

> Features especially good illustrations of flower parts.

Oechsli, Helen, and Kelly. *In My Garden*. New York: Macmillan Publishing Co., 1985.

> A child's step-by-step guide to planting and maintaining a garden. Includes techniques for thinning and transplanting, weeding and composting.

Schnieper, Claudia. *An Apple Tree through the Year*. Minneapolis: Carolrhoda Books, Inc., 1987.

> Full-color photographs and text explaining the yearly cycle of growth and fruit development.

Suzuki, David. *Looking at Plants*. New York: Warner Books, 1985.

> A book of simple plant projects for both indoors and out.

Thompson. *Trees*. London: Usborne-Hayes Publishing Ltd., 1980.

> Well-illustrated book featuring a wide variety of trees.

Wexler, Jerome. *Flowers, Fruits and Seeds.* New York: Prentice Hall, 1988.

Wyler, Rose. *Science Fun with Peanuts and Popcorn.* New York: Julian Messner, 1986.

> Simple projects to explore germination, growth, and development. Fun activities are included, such as tongue twisters and riddles.

Franchere, Ruth. *Cesar Chavez*. New York: Harper and Row, 1970.

> Tells how Chavez organized farm workers to protest against unfair treatment.

Mitchell, Barbara. *Pocketful of Goobers*. Minneapolis: Carolrhoda Books, Inc., 1986.

Moore, Eve. *The Story of George Washington Carver*. New York: Scholastic, Inc., 1971.

> Both books tell the story of how Carver, a black man, overcame enormous obstacles to become an outstanding plant scientist. He invented hundreds of uses for the peanut.

Computer Programs

"MECC Graph" and "MECC LUNAR GREENHOUSE." Available from: MECC, 3490 Lexington Ave. N., St. Paul, Minnesota 55126.

> Apple II series, printer recommended. Children collect data, enter information, and generate graphs. For grades 7 to 9, but easily adapted for younger children.

"Project Zoo: Adventures with Charts and Graphs." Available from: National Geographic Society, Educational Services, Department 89, Washington, D.C. 20036.

> Apple II series, color monitor and printer recommended but optional. Contains three disks, filmstrip, audiocassette, manual, activity sheets, zoo fact book, pretest and post-test. For grades 3 to 5.

Videos

"Pollination." Available from: National Geographic Society.

> Illustrates flower anatomy, pollination, and fertilization. Outstanding footage of ordinary as well as exotic pollinators. For grades 4 to 9; 23 minutes.

Filmstrips and Transparencies

"Insects: How They Help Us." Available from: National Geographic Society.

> Emphasizes the beneficial aspects of insect life. Comes with cassette. For intermediate grades; 15 minutes.

"How Living Things Depend on Each Other." Available from: National Geographic Society.

> Designed to teach the interdependence of living things. Comes with cassette. For intermediate grades; 13 minutes.

"Honeybee II Symbiosis with Flowering Plants." (Overhead Transparency.) Available from: Carolina Biological Supply Co., Burlington, North Carolina 27215.

> Beautifully drawn and colored, the transparencies show the honeybee anatomy relative to the *Brassica* plant. Advanced, but easily adapted for younger children.

APPENDIX E

Life Cycle Cards

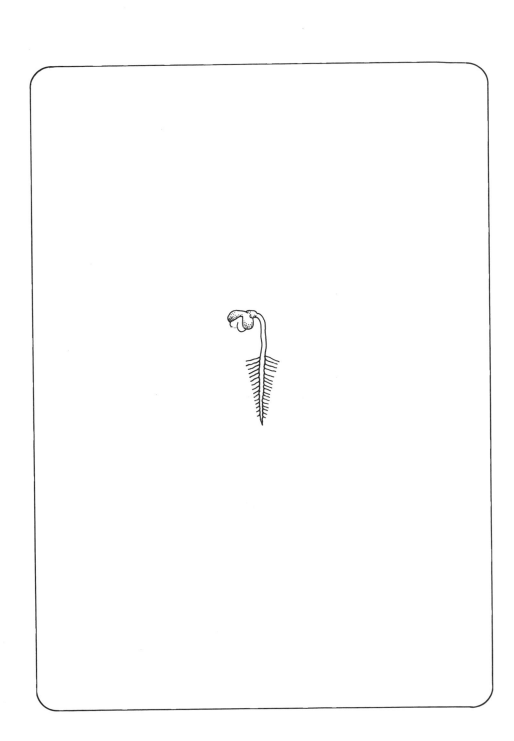

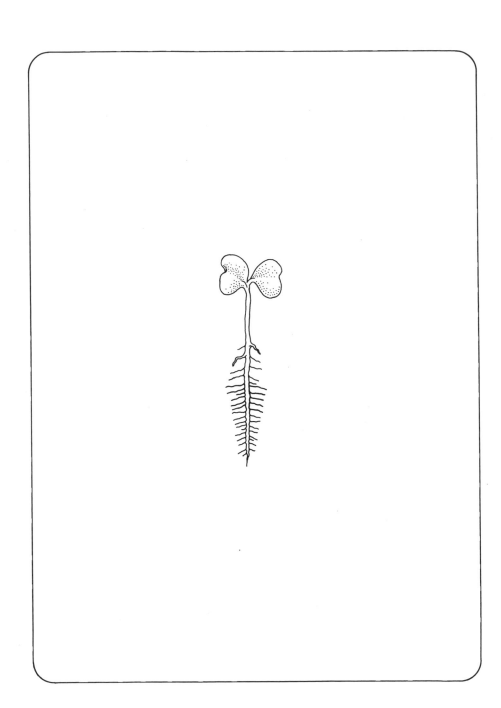

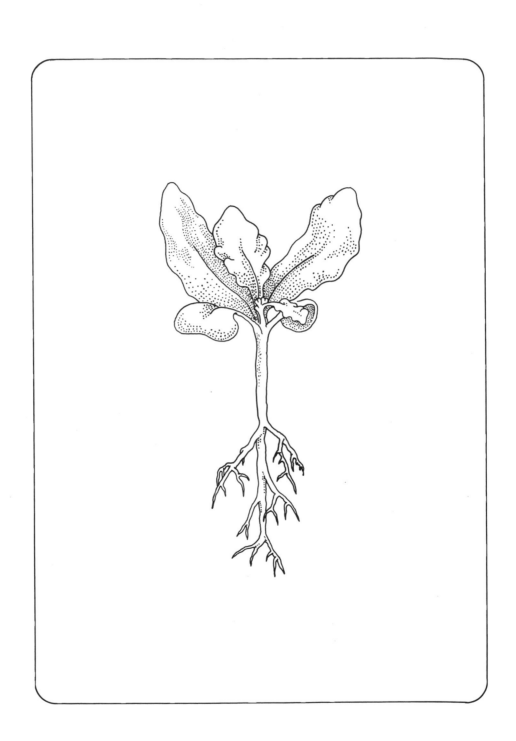

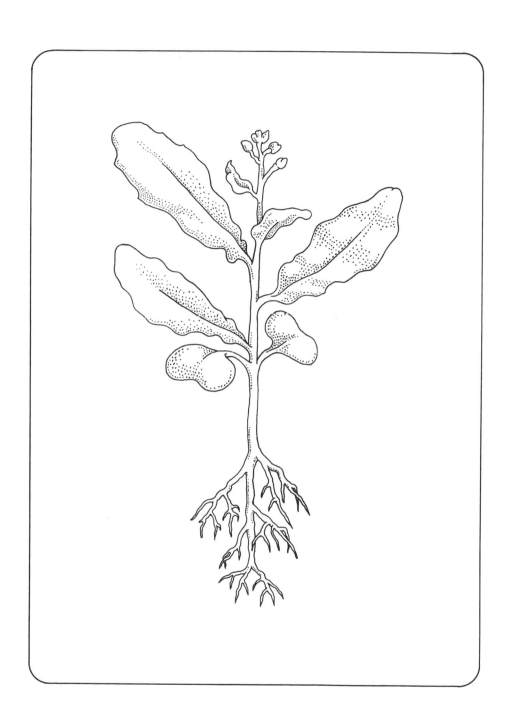

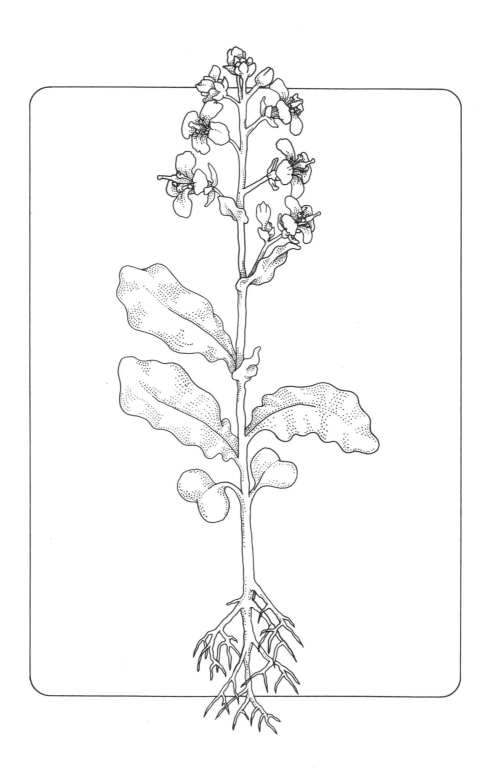

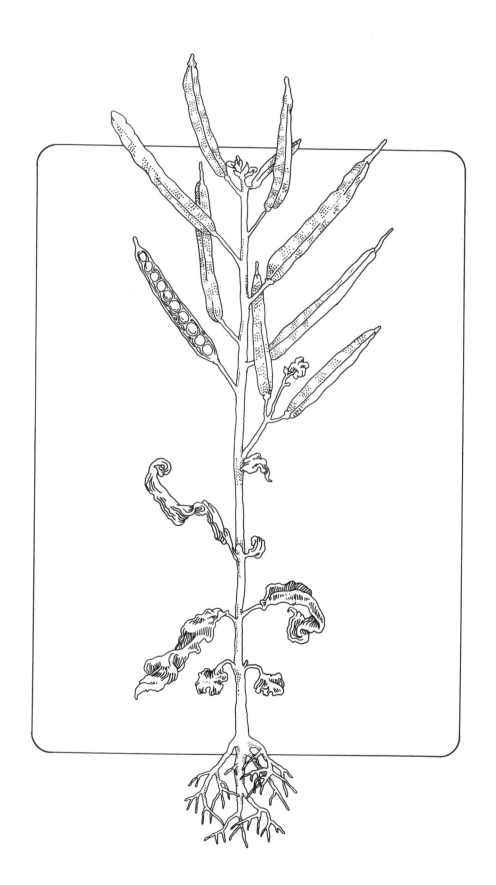

Black Line Masters

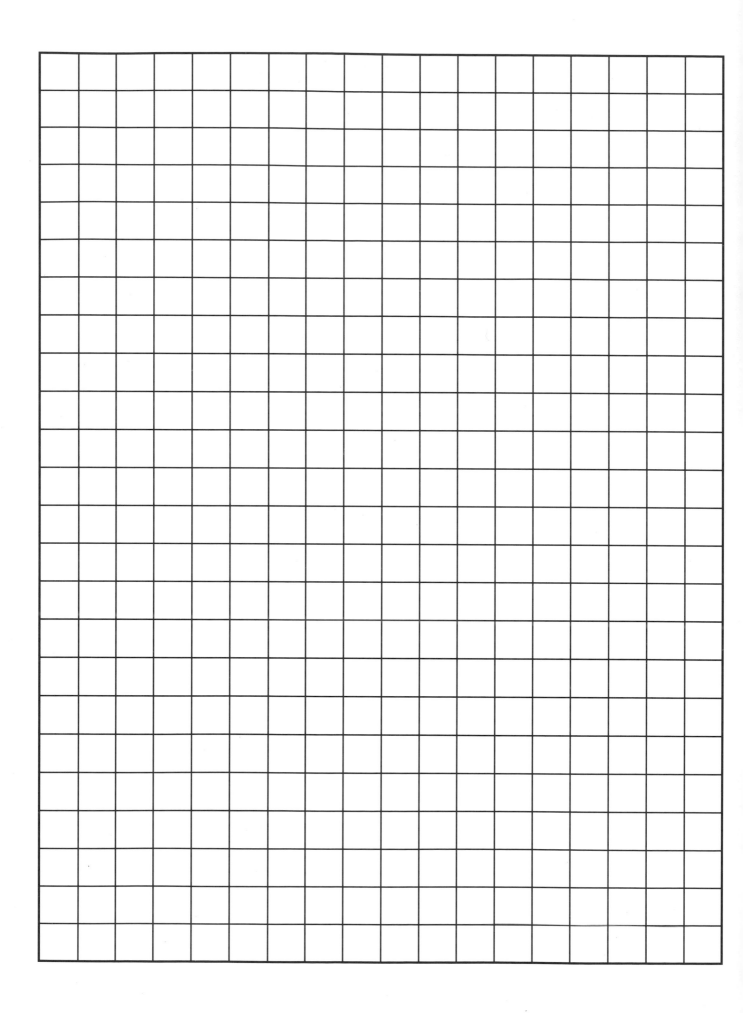

Wisconsin Fast Plants™ Growth Graph

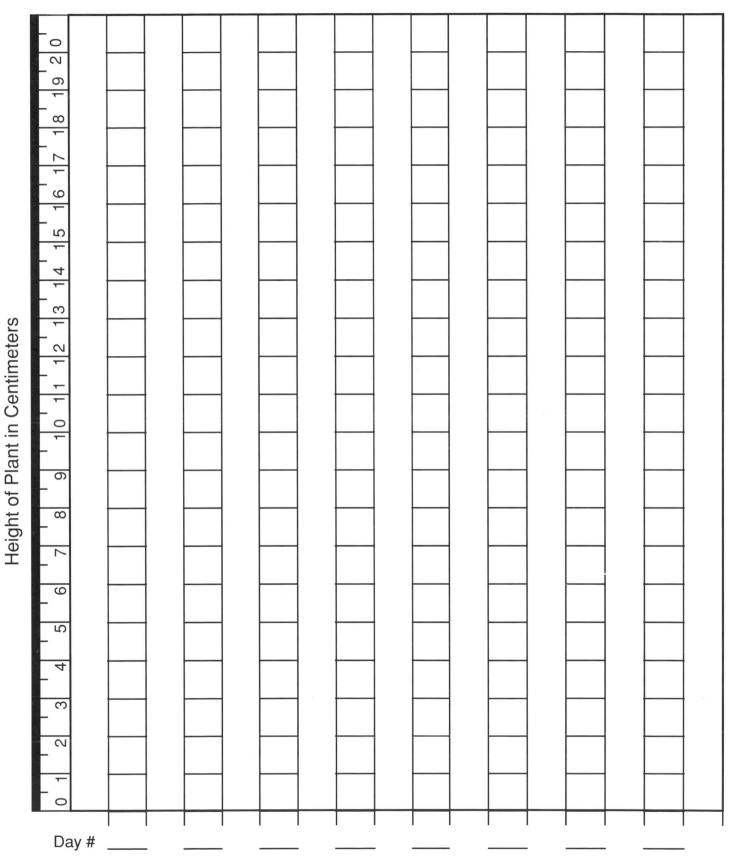

Height of Plant in Centimeters

Day # ____ ____ ____ ____ ____ ____ ____ ____ ____ ____

Age of Plant in Days

Materials Reorder Information

During the course of hands-on science activities, some of the materials are used up. The consumable materials from each Science and Technology for Children™ unit can be reordered as a unit refurbishment set. In addition, a unit's components can be ordered separately.

For information on refurbishing *Plant Growth and Development* or purchasing additional components, please call Carolina Biological Supply Company at **800-334-5551.**

National Science Resources Center Advisory Board

Chairman

Joseph A. Miller, Jr., Chief Technology Officer and Senior Vice President for Research and Development, E. I. du Pont de Nemours and Company, Wilmington, DE

Members

Ann Bay, Director, Office of Elementary and Secondary Education, Smithsonian Institution, Washington, DC

DeAnna Banks Beane, Project Director, YouthALIVE, Association of Science-Technology Centers, Washington, DC

Fred P. Corson, Vice President and Director, Research and Development, The Dow Chemical Company, Midland, MI

Goéry Delacôte, Executive Director, The Exploratorium, San Francisco, CA

JoAnn E. DeMaria, Elementary School Teacher, Hutchison Elementary School, Herndon, VA

Peter Dow, Director of Education, Buffalo Museum of Science, Buffalo, NY

Hubert M. Dyasi, Director, The Workshop Center, City College School of Education (The City University of New York), New York, NY

Bernard S. Finn, Curator, Division of Information Technology and Society, National Museum of American History, Smithsonian Institution, Washington, DC

Robert M. Fitch, President, Fitch & Associates, Taos, NM

Jerry P. Gollub, John and Barbara Bush Professor in the Natural Sciences, Haverford College, Haverford, PA

Ana M. Guzmán, Vice President, Cypress Creek Campus and Institutional Campus Development, Austin Community College, Cedar Park, TX

Anders Hedberg, Director, Center for Science Education, Bristol-Myers Squibb Pharmaceutical Research Institute, Princeton, NJ

Richard Hinman, Senior Vice President (retired), Research and Development, Pfizer Inc., Groton, CT

David Jenkins, Associate Director for Interpretive Programs, National Zoological Park, Smithsonian Institution, Washington, DC

Mildred E. Jones, Educational Consultant, Baldwin, NY

John W. Layman, Director, Science Teaching Center, and Professor, Departments of Education and Physics, University of Maryland, College Park, MD

Leon Lederman, Chairman, Board of Trustees, Teachers Academy for Mathematics and Science, Chicago, IL, and Director Emeritus, Fermi National Accelerator Laboratory, Batavia, IL

Sarah A. Lindsey, Science Coordinator, Midland Public Schools, Midland, MI

Lynn Margulis, Distinguished University Professor, Department of Botany, University of Massachusetts, Amherst, MA

Theodore Maxwell, Senior Advisor for Science, National Air and Space Museum, Smithsonian Institution, Washington, DC

Mara Mayor, Director, The Smithsonian Associates, Smithsonian Institution, Washington, DC

John A. Moore, Professor Emeritus, Department of Biology, University of California, Riverside, CA

Carlo Parravano, Director, Merck Institute for Science Education, Rahway, NJ

Robert W. Ridky, Professor of Geology, University of Maryland, College Park, MD

Ruth O. Selig, Executive Officer for Programs, Office of the Provost, Smithsonian Institution, Washington, DC

Maxine F. Singer, President, Carnegie Institution of Washington, Washington, DC

Robert D. Sullivan, Assistant Director for Public Programs, National Museum of Natural History, Smithsonian Institution, Washington, DC

Gerald F. Wheeler, Executive Director, National Science Teachers Association, Arlington, VA

Richard L. White, Executive Vice President, Bayer Corporation, Pittsburgh, PA, and President of Fibers, Organics, and Rubber Division, and President and Chief Executive Officer, Bayer Rubber Inc., Canada

Paul H. Williams, Atwood Professor, Department of Plant Pathology, University of Wisconsin, Madison, WI

Karen L. Worth, Faculty, Wheelock College, and Senior Associate, Urban Elementary Science Project, Education Development Center, Newton, MA

Ex Officio Members

Rodger Bybee, Executive Director, Center for Science, Mathematics, and Engineering Education, National Research Council, Washington, DC

E. William Colglazier, Executive Officer, National Academy of Sciences, Washington, DC

J. Dennis O'Connor, Provost, Smithsonian Institution, Washington, DC

Barbara Schneider, Executive Assistant for Programs, Office of the Provost, Smithsonian Institution, Washington, DC